T0235438

SpringerBriefs in Environment, Security, Development and Peace

Volume 33

Series Editor

Hans Günter Brauch, Sicherheitspolitik, Peace Research & European Security Studies, Mosbach, Baden-Württemberg, Germany

More information about this series at http://www.springer.com/series/10357
http://www.afes-press-books.de/html/SpringerBriefs_ESDP.htm
http://www.afes-press-books.de/html_ESDP1.htm

Yusniza Kamarulzaman · Farrah Dina Yusop ·
Noorhidawati Abdullah · Azian Madun ·
Kwan-Hoong Ng

Public Perceptions of Radiation Effects on Health Risks and Well-Being

A Case of RFEMF Risk Perceptions in Malaysia

 Springer

Yusniza Kamarulzaman
Faculty of Business and Accountancy
University of Malaya
Kuala Lumpur, Malaysia

Noorhidawati Abdullah
Faculty of Computer Science
and Information Technology
University of Malaya
Kuala Lumpur, Malaysia

Kwan-Hoong Ng
Faculty of Medicine
University of Malaya
Kuala Lumpur, Malaysia

Farrah Dina Yusop
Faculty of Education
University of Malaya
Kuala Lumpur, Malaysia

Azian Madun
Academy of Islamic Studies
University of Malaya
Kuala Lumpur, Malaysia

ISSN 2193-3162   ISSN 2193-3170 (electronic)
SpringerBriefs in Environment, Security, Development and Peace
ISBN 978-981-32-9893-4   ISBN 978-981-32-9894-1 (eBook)
https://doi.org/10.1007/978-981-32-9894-1

This Springer imprint is published by the registered company Springer Nature Singapore Pte Ltd.
The registered company address is: 152 Beach Road, #21-01/04 Gateway East, Singapore 189721, Singapore

Preface

Risk perceptions play a pivotal role in health behaviour theories. Understanding the impact of perceived risk on well-being is essential in order to investigate the explanatory value and effectiveness of the interventions influencing these beliefs. The interrelationships between risk knowledge, attitude and behaviour among different groups of the general public are very complex such that the generalization of relationship cannot easily be concluded. Investigating these relationships is essential in order to identify and predict the behaviour of the general public regarding health issues. A better understanding of the influence of this knowledge on attitudes and risk reduction in a particular culture or country could significantly help in designing effective educational/awareness programmes and health messages.

During the past few years, the mainstream media such as the television news channels and newspapers have sporadically reported the fear and concern among the general public with regard to the presence of telecommunication towers in many residential areas. This is due to the perception of some quarters of the general public that these towers radiate electromagnetic signals, which may pose hazard and risk to human health. Therefore, this study was commissioned as part of the Government's effort to identify and examine the perception of risk among the general public on this matter.

Since the telecommunication towers are built for the purpose of transmitting electronic signals which have the properties of electromagnetic fields, this book is about the study on the investigation of the public perception on the effects of radio-frequency electromagnetic field (RFEMF) on human health and well-being. This study seeks to expand knowledge and understanding of different risk perceptions related to the radiation of RFEMF, which would explain the gap in the literature regarding the relationship between risk perceptions that lead to public behaviours.

This book presents empirical findings of a national study that unveils two key factors affecting public risk perceptions: psychographic and personal factors. It brings a more collective and cultural understanding of public perceptions of radiation risks via systematic mixed-method research approach. While the radiation risk is recognizable and unavoidable, the relevant stakeholders should be more proactive and committed to communicate and rectify the perception of radiation. This book can be a source of reference to understand the debate and to invite more partici-patory dialogues on radiation risk perceptions among the public.

Kuala Lumpur, Malaysia   Yusniza Kamarulzaman
                                                                          Farrah Dina Yusop
                                                                          Noorhidawati Abdullah
                                                                          Azian Madun
                                                                          Kwan-Hoong Ng

Acknowledgements We would like to acknowledge the Malaysian Communications and Multimedia Commission (MCMC) and University of Malaya who supported this research work under the Grant No. 5302-03-1091.

Contents

List of Figures

List of Tables

Chapter 1
Introduction to Radiation and Risk Perception

A. Madun

1.1 Radiation and Radio Frequency Electromagnetic Fields

1.1.1 Background of Radio Frequency Electromagnetic Fields

Human beings' understanding about magnetic, electric and electromagnetic fields has been growing and developing through various stages since 1600 with the work of scientists such as Isaac Newton, Michael Faraday and James Maxwell. Electromagnetic fields were first discovered in the 19th century when scientists noticed that electric arc could be reproduced at a distance, with no connecting wires in between. This led scientists to believe that the sparks could be used as signals to communicate over long distances without wires (Forbes & Mahon 2014). This was, in fact, the beginning of the development of mobile technologies or wireless devices.

Electromagnetic fields are typically generated by an alternating current in electrical conductors. At the low extreme, the frequency of the alternating current is one cycle in thousands of years, while it can go to trillions or quadrillions of cycles per second at the high extreme. The frequency of the electromagnetic field is usually expressed in terms of a unit called hertz with an abbreviation of Hz. One Hz equals one cycle per second and one megahertz (MHz) equals one million cycles per second. In short, the electromagnetic field consists of waves of electric and magnetic energy moving together through space at the speed of light (Purcell 2012).

RFEMF is useful in everyday life. The most important use of RFEMF is in providing telecommunications services such as radio and television broadcasting, cellular phones, personal communications services, pagers, cordless telephones and radio communications for police and fire departments as well as amateur radio.

RFEMF can also be used for non-communication purposes such as radar and heating. Radar is a valuable tool used by traffic speed enforcement, air traffic control at the airports as well as military surveillance. An intense level of RFEMF radiation

© The Author(s), under exclusive license to Springer Nature Singapore Pte Ltd 2020
Y. Kamarulzaman et al., *Public Perceptions of Radiation Effects on Health Risks and Well-Being*, SpringerBriefs in Environment, Security, Development and Peace 33, https://doi.org/10.1007/978-981-32-9894-1_1

can produce heat that can be used to rapidly heat materials used in many industries, including moulding plastic materials, wood products and sealing items. It is also used in medical applications such as magnetic resonance imaging (Jin 1999).

1.1.2 Basis of Concerns of RFEMF

The concern related to electromagnetic field arises when it interacts with organic tissue, consequently health-related concerns arise. One example of the application of electromagnetic field is X-ray machines which are commonly used in the hospitals. An interaction with the electromagnetic field can lead to "ionization". Ionization is a process where electrons are stripped from atoms that can lead to damage in biological tissue, including genetic material of living organisms. X-ray is an example of an electromagnetic field which is sufficiently high to ionize biological material such that the X-ray is capable of detecting broken bones, tumours, dental decay and abnormalities within the body without invasive surgery or causing any pain (Kimura et al. 2008).

Ionizing radiation may possibly cause different types of cancer such as stomach, liver, colon, lung, breast, uterine and thyroid cancer and leukaemia as well as genetic effects. High radiation exposures can also damage blood and tissues, including the heart, eyes, intestines, skin, and reproductive organs, depending on the type of radiation (alpha, beta, gamma), the route of absorption and its potency (Cember & Johnson 2009).

Non-ionizing radiation is also electromagnetic radiation but it has less energy. Therefore, it is not powerful enough to cause the ionization (removal of electrons) of molecules. It includes many types of electromagnetic radiation ranging in energy from extremely low frequency radiation to ultraviolet radiation. Hospitals employ equipment that generates many types of non-ionizing radiation such as magnetic resonance imaging (MRI) and lasers. The health effects of non-ionizing radiation are expected to be related to many factors such as periods of exposure, wavelength or change in electric and magnetic fields over time and space. As a result, workers in hospitals should be aware of the non-ionizing and ionizing hazards that they are facing in their job (Gorman et al. 2013).

Therefore, RFEMF is often related to "radiation". However, the term radiation associated with RFEMF is often confused with the term "radioactive". Hence, radiation is always perceived to do with a bad connotation. Radiation in RFEMF refers to the motion of the RFEMF through space that may cause heating, whereas radioactive refers to the decaying process of an unstable atom that loses energy by emitting radiation. Radiation related to RFEMF may or may not pose danger to human depending on the levels of radiation exposure, while radiation related to radioactive contains substances that are ultimately hazardous to human. This might explain why there is a bad perception with regards to RFEMF due to its association with the term radiation.

There are three common ways to measure RFEMF radiation; firstly, volt per meter which is used to express the strength of the electric field; secondly, watts per square meter which is used to measure the density of exposure; and thirdly, watts per kilogram which is used to measure the quantity of energy absorbed in a body widely known as the "Specific Absorption Rate" or "SAR" (Jin 1999).

An exposure to very high levels of RFEMF can be detrimental due to its ability to heat biological tissue rapidly, which in turn increases body temperature. This can cause tissue damage owing to the body's inability to cope with such a high body temperature or because of dissipating the excessive heat. Two areas of the body, the eyes and the testes, are particularly vulnerable because of the relative lack of available blood flow to dissipate the excess heat load.

At relatively low levels of exposure to RFEMF, the heat produced is insignificant and the evidence does not suggest any harmful biological effects. Nonetheless, standards-setting organisations and government agencies continue to monitor the latest experimental findings to confirm their validity and determine whether changes in safety limits are needed in order to protect human health.

In short, although there are no clear evidence to suggest a link between electromagnetic fields and negative health effects on humans, considerable prudence is necessary such as persons who have metal-containing implants fitted in their body to avoid superconducting magnet (Perrin & Souques 2012). Similarly, people who are exposed to electromagnetic radiation at the workplace should avoid long exposure to the radiation or reduce the exposure to high frequency (Gorman et al. 2013).

1.1.3 Differences of Opinion Among Experts

Electromagnetic fields are most diffuse and ubiquitous, especially because many consumer goods and new advanced technological devices are being developed based on the application of electromagnetic fields. As a result, these products can be categorised by the order of frequencies of electromagnetic spectrum and the energies developed in the range of frequencies. For example, the electric power lines are categorised under low frequency, while X-ray machine belongs to high frequency (Perrin & Souques 2012) as shown in Fig. 1.1.

As the application of electromagnetic fields increases in consumer goods, there is an increase in the research on the possible health and biological effects of RFEMF that have been carried out in Europe, North America, Japan and other countries. These research activities are supported by public and private funding bodies at both the national and international levels (EFHRAN 2012).

The extent and diversity of these activities, encompassing many areas of medical and biological research, as well as the latest developments in physics and engineering make it particularly difficult to provide relevant, authoritative and timely input for the development of public health policies. Furthermore, it is possible that specific assessments for one situation can be misinterpreted or inappropriately applied to other conditions.

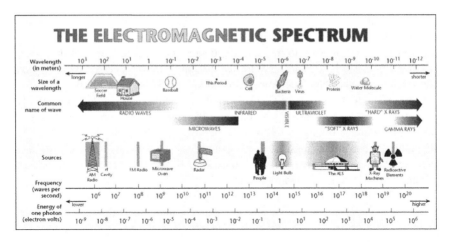

Fig. 1.1 Electromagnetic Fields Spectrum - Range of Wavelengths.
Source https://www2.lbl.gov/MicroWorlds/ALSTool/EMSpec/EMSpec2.html

The World Health Organisation (WHO) has issued continuous statements from time to time with regard to the risk of RFEMF as the usage of mobile phone becomes universally accepted. As the transmission of a mobile phone communication signal is one of the sources of RFEMF, in their latest statement, they are of the opinion that there are no adverse health effects that could have been established due to the use of mobile phone (World Health Organization 2014). Their opinion is based on the report by the International Commission on Non-Ionizing Radiation Protection (ICNIRP), which is also supported by researchers in Europe who have examined the literature on scientific evidence on the risk of RFEMF. They are also of the opinion that the health risk due to human exposure to electromagnetic fields is either limited or inadequate (EFHRAN 2012).

However, there are differences of opinion among experts regarding the risks of RFEMF as differences arise from various perspectives. For example, Starkey (2016) argued that the official assessment of the safety of RFEMF is inaccurate because the executive summary and overall conclusions provided by the government agency did not accurately reflect the scientific evidence available. Hence, he suggested the presence of a conflict of interest in the advisory group. Hardell (2017) claimed the presence of a conflict of interest among advisory group members of the WHO on the safety of RFEMF. This is because five of the six members are industry-affiliated, resulting in the dismissal of adverse health effects of RFEMF radiation.

Nonetheless, there were experiments that have been performed to investigate the impact of RFEMF. For example, Akdag et al. (2016) found that prolonged exposure to 2.4 GHz frequency of Wi-Fi device can significantly affect some organ tissues of rats as compared to rats that were not exposed to Wi-Fi.

In conclusion, the risk of RFEMF full of contentious issues from opponents and proponents is debatable. Both sides have their own arguments and supporting evidence. Hence, it is important to investigate public perceptions of the risk of RFEMF and its implication.

1.2 Risk Perception

1.2.1 Background of Risk Perception

Risk is defined as a combination of the probability or frequency of damage, injury, liability, loss or hazard caused by external or internal vulnerabilities that may be avoided through pre-emptive action (Harding 1998). Risk is also associated with unfortunate events such as death, injuries and damages, so the significance of events indicates the strength of the signals of risk (Slovic 1987).

This notion usually provides the basis for the formal risk assessment and is the form in which risk is often conceptualised by the technical experts. While this view considers risks primarily quantitatively and typically rejects other forms of risk perception as irrational, laypeople generally have a more intuitive, qualitative risk concept that is not limited to the probability or to the extent of damage (Renn & Levine 1991).

People perceive risk differently depending on the knowledge people have about the risk which contributes to the differences in perception of risks between the experts and the laypeople (Slovic 1987). Laypeople usually associate images they have in mind with risks. For example, if they identify mobile base station with negative images, their perception of risks on the mobile base station is high. On the other hand, if they have a positive image associated with the mobile base station, their perceived risk is lower (Dohle et al. 2012).

This led many researchers to propose that risk is socially constructed or perceived rather than an objective state of nature. Risk perception is frequently held to be crucial in the understanding and management of risk in policy contexts (Sjöberg 1987). Before adopting or accepting new invention, People usually weigh both the benefits and risks simultaneously (Davis 1989).

Risk perception is an important element in policy development and public decision-making. It remains one of the most important challenges in nuclear risk governance, as proven by the lessons learnt from past events. Understanding and managing risk perception is crucial in any organisation (Sjoberg 2000). The inability to gauge the public perceptions of a controversial project will affect the continuity of the project (Slovic 2010).

As new knowledge emerged, the risks and benefits perceived by people may change over time. The benefits are usually examined from an economic efficiency perspective. However, some risks are difficult to determine in terms of finances such as risks to human health and the environment (Heath et al. 1995). In addition, the perceived benefits can also be categorised as favourable or unfavourable. If people expect a more favourable benefit, then the perceived risk is low. On the other hand, when the benefit is unfavourable, the perceived risk is high (Alhakami & Slovic 1994).

1.2.2 The Determinants of Risk Perception

Although it is hard to understand, the risk perception has always been the focus of policymakers and researchers. According to Sjoberg (2000), there are various ways to explain risk perception which can be categorised as follows: technical estimates (real risk), heuristics (human biased), psychometric model (scales as measurements of risk) and cultural theory (categorising people based on their hazard concern).

While there is no model that can be adapted to all the phenomena, generally people can accept a certain level of risk in their lives, if necessary, to achieve certain benefits. Similarly, people are also likely to accept the risk if the benefit is higher than the risk (Slovic 2000). Individuals often make calculated risks. If they do not know or understand the situation or do not have the experience to make an informed decision, they may choose to take the calculated risk.

There are many factors that may influence human beings' decision-making which will eventually affect their behaviour (Hillson 2004). However, studies on the relationship between risk perception and behaviours have revealed mixed results. The typical theoretical prediction is that the risk perception is negatively correlated with risk-taking behaviour. It means that people behave in such a way that they can protect themselves from the risk, i.e. protective behaviour (Reyna & Farley 2006). However, there are studies which indicate positive correlation between risk perception and behaviour. It means that the risk does not hinder people from engaging in what they were doing before, i.e. reflective behaviour (Johnson et al. 2002).

Essentially, this shows that people are willing to accept a certain level of risk in their lives, if necessary, to achieve certain benefits. The higher the benefit, the more likely they will accept the risk (Slovic 2000). Individuals often make calculated risks. If they do not know or understand the situation or do not have an experience base to make informed decisions, they may choose to take a calculated risk. If people take safety for granted, they may not stop to consider the whole picture. Sometimes, they are influenced by the people around them to accept risks that they normally would not.

On the other hand, the risk perception does not just apply to individuals; it also applies when people work in groups or teams. People respond to their assessment of circumstances and what their peers tell them. If a team member, whom they respect and whom they believe is more experienced than them, tells them that something is safe, they tend to accept that respected member's decision. If a person of authority deems an environment or piece of equipment to be safe, they generally do not question that person's conclusion (Vaughan & Hogg 1998; Geller 2005). Most groups have a natural leader who sets the group culture, is never questioned and always has the final say.

Besides the trade-off between benefits and loss, risk perception is also related to intuitive beliefs. For example, Wiedmann et al. (2017) found that the general public tends to believe that many people are highly exposed to the risks of RFEMF although they do not know much about the health impact of RFEMF. They suggested that this could be due to the inaccurate method of measuring the risk perception which is biased towards the exaggeration of public views.

In summary, it is not easy to generalise risk perception in any field of study. Sometimes, the risks are exaggerated as the media compete for public attention and made the issue driven by public sentiment rather than the actual facts (Kahneman 2012). Therefore, risk perception is a highly personal process of decision making based on an individual's frame of reference developed over a lifetime, among many other factors (Brown 2014).

1.3 Risk Communication

Risk communication is basically to communicate the danger of certain thing/issue to a specific audience or general public so that the audience can make an informed decision. Risk communication usually involves two-way communication between the organization that is managing the risk and the audience carrying a dialogue (Lundgren & McMakin 2018).

Risk communication is important as it helps reduce the risks such as the spreading of disease which will eventually save lives and protect national and local economies (World Health Organization 2015). By properly communicating the risk, public anxiety can be reduced, controversy can be dispelled and the conflicts can be avoided (Wiedeman 1999; World Health Organization 2002). Too often misleading information is transmitted via word of mouth, propagation of news via internet and media attention. With effective risk communication involving stakeholders such as individuals, community-based organisations and non-government agencies, the public would be more informed of the actual situation. In this case, risk communication serves to explain scientific uncertainty, as the public usually interprets uncertainty as a declaration of the existence of real risks (WHO 2002).

Gray (1998) presented several arguments which call for risk communication. Firstly, people living in a democracy deserve to have input into decisions that affect them. As the people have the right to be heard and to receive information, certain approach for communication is required. Secondly, despite the initial effort to enter in a dialogue with the affected people, risk communication would help improve mutual understanding and trust over the long term. Thirdly, risk communication could lead to better decisions as it provides additional knowledge such as local knowledge, possible biases and hidden assumptions. Ultimately, based on the collective information gained by the decision makers, risk communication will lead to improved policy-making (Qiu et al. 2016).

WHO (2002) has identified eight groups of key stakeholders concerning the RFEMF issue; scientific community, health and legal practitioners, industry players, non-governmental organisations, media, general public and government. All key stakeholders have different roles in their respective communities.

For example, the scientific community should be independent and apolitical and is responsible for providing technical information which could help the public understand the benefits and risks of EMF. The health practitioners are responsible to provide input from the health aspect concerning the EMF issue. The legal practitioners are responsible for legal standing and involvement in the issue under question in order to ensure that everyone involved is corresponding to the dictates of the law (National Research Council 1989).

The industry, despite being viewed acting based on profit motive by the public, should be active in managing risk and initiating open communication with the public. Associations which may include environmental groups, community-based organisations and non-government agencies are important in making the public sentiment heard by other stakeholders. Media which may include the internet, television, newspaper and radio could serve effectively in increasing problem awareness and broadcast information, but it could also be effective in disseminating incorrect information. The general public which includes landowners and concerned citizens might be the greatest determinant to the success or failure of a proposed technology project. However, they should be careful of misleading information caused by media. The government officials are important to devise standards and guidelines, which should be based upon information from other major stakeholders (WHO 2002).

1.3.1 The Benefits of Risk Communication

Given the importance of risk communication, an important question arises: What are the determinants for risk communication? Peters et al. (1997) emphasised trust and credibility as the key determinants for successful risk communication. More specifically, their research had found three most important factors that influence public confidence and trust for risk communication: (1) perception of knowledge and expertise, (2) perception of openness and honesty and (3) perception of concern and care. They concluded by suggesting the following to improve risk communication:

(1) For the industry, to increase the public perception of concern and care.
(2) For the government, to increase the public perceptions of commitment.
(3) For the citizen, to increase public perceptions of knowledge and expertise.

Poor risk communication would cause the public to have insufficient information they need to protect themselves and their families, which could ultimately lead to massive panic among the public (World Health Organization 2015) such as the severe acute respiratory syndrome (SARS) and avian flu outbreaks. The delayed communication caused SARS early symptoms to be largely unnoticed as many clinicians were unaware of this epidemic threat (Qiu et al. 2016). This eventually caused the significant epidemic break-out that gained considerable strength before it was recognised by the government.

Ineffective risk communication could also devastate communities and economy if it leads to illness, fear and death (World Health Organization 2015). For instance, in the SARS outbreak crisis, China's delayed detection and poor communication eventually caused civil unrest which damaged their economy and reputation (Qiu et al. 2016). In the long run, poor risk communication would also cause the government to lose the public's trust, especially if they are perceived to hide information or release contradictory information.

Numerous studies have highlighted the importance of effective risk communication in enabling people to make informed choices and participate in deciding how risks should be managed. Effective risk communication provides people with timely, accurate, clear, objective, consistent, and complete risk information. It is the starting point for creating an informed population that is involved, interested, reasonable, thoughtful, solution-oriented, cooperative and appropriately concerned. The U.S. National Academy of Sciences (National Research Council 1989) has defined risk communication as "an interactive process of exchange of information and opinion among individuals, groups and institutions." Risk communication is the two-way exchange of information about risks, including the risks associated with radiation and radiological events and emergencies. Policymakers must recognise the importance of risk and crisis communication planning (Covello & Sandman 2001). To communicate risks effectively during and after an emergency, Covello strongly advocated an APP approach: (A) anticipation, (P) preparation and (P) practice.

In summary, risk communication should be properly planned and the planning should be comprehensive. The organisation managing the risk must determine the objectives of the risk communication efforts, identify the target audience, understand the legal mandate related to the risk and the constraints in undertaking the effort including ethical issues, determine the appropriate methods of communication and finally evaluate the risk communication that has been undertaken (Lundgren & McMakin 2018).

1.3.2   The Amplification of Risk

One of the most perplexing problems in risk analysis is why some relatively minor risks or risk events, as assessed by technical experts, often elicit strong public concerns and result in substantial impacts upon society and economy. It shows that the social experience of risk is not confined to the technical definition of risk, i.e., the product of probability and magnitude. What human beings perceive as a threat to their wellbeing and how they evaluate probabilities and magnitudes of unwanted

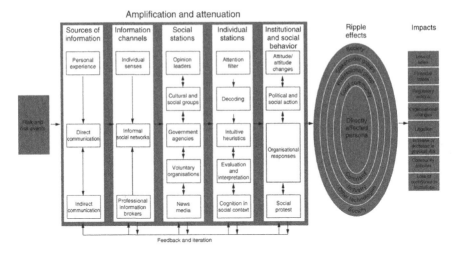

Fig. 1.2 The Social Amplification of Risk Framework (SARF).
Source Pidgeon & Henwood (2010)

consequences is determined by values, attitudes, social influences and cultural identity (Tversky & Kahneman 1974).

The interaction of psychological, social, institutional and cultural processes with risk may amplify or attenuate public responses to the risk or event. A structural description of the social amplification of risk is shown in Fig. 1.2. The social amplification of risk (SARF) combines theories in the fields of psychology, sociology, anthropology and communications. As shown in Fig. 1.2, the amplification of risk occurs at two stages: during the transfer of information about the risk and in the response mechanisms of society. The amplified risk leads to behavioural responses, which, in turn, result in secondary impacts (Kasperson et al. 1988).

The framework endeavours to explain the process by which risks are amplified, receiving public attention, or attenuated, receiving less public attention. All links in the communication chain (individuals, groups, media etc.) contain filters through which information is sorted and understood. The framework may be used to compare responses from different groups in a single event or analyse the same risk issue in multiple events. In a single risk event, some groups may amplify their perception of risks while other groups may attenuate their perceptions of risk.

The main idea of SARF states that risk events interact with individual psychological, social and other cultural factors in ways that either increase or decrease public perceptions of risk. Behaviours of individuals and groups then generate secondary social or economic impacts besides increasing or decreasing the physical risk itself (Kasperson et al. 1988).

These ripple effects caused by the amplification of risk include enduring mental perceptions, impacts on business sales, change in residential property values and changes in training and education or social disorder. These secondary changes are perceived and reacted to by individuals and groups resulting in third-order impacts.

As each higher-order impact is reacted to, they may ripple to other parties and locations. Traditional risk analyses neglect these ripple effect impacts and thus greatly underestimate the adverse effects from certain risk events. Public distortion of risk signals provides a corrective mechanism by which society assesses a fuller determination of the risk and its impacts to such things not traditionally factored into a risk analysis (Kasperson & Roger 2005).

1.3.3 Trust in the Government

As described earlier, the technology that is derived from RFEMF is mainly applied to consumer goods which require Government approval before they can be sold to the general public. Similarly, the construction of telecommunication transmission towers requires permission from the federal and local government. It means that the study of RFEMF would also have to involve the Government who was also one of the stakeholders that have the authority in giving approval. In other words, the study would be more comprehensive by examining the authority of the Government pertaining to RFEMF.

Previous studies show that the public trust in the Government is related to the performance of the Government, which indicates that as long as the public is satisfied with the performance, they will put their trust in the Government (Yang & Holzer 2006; Torcal 2014). Besides performance, the transparency of the Government also plays a role in increasing public trust in Government. With the prevalence of e-government, the general public can have access to timely and reliable information on the Government's procedures and enable them to evaluate the decision making process. It increases the sense of democracy among the general public and empowers them to monitor the Government performance; hence restoring high confidence in the Government (Pina et al. 2007; Kim & Lee 2012).

In short, the communication of risk would be more effective when the general public is satisfied with the performance of Government that are transparent and not corrupted.

1.4 About the Study

1.4.1 Background

As early as 2005, there were many newspaper reports regarding the danger of mobile phone towers or bases that were erected in residential areas. For example, in April 2008, a tabloid newspaper reported about a claim that mobile phone tower could lead to cancer to the people living nearby due to the radiation from the tower. Recently, another newspaper report refuted the claim by saying that the telecommunication tower does not pose any risk to human health as the reading of the exposure was lower than suggested by the authority (Utusan Online 2018). Such newspaper coverage may cause discomfort and disturb the peace of mind among some quarters of the general public. Hence, this study was commissioned by the Government of Malaysia as part of the study to examine the impact of telecommunication towers on health (Utusan Online 2008).

Amid the contradicting reports from various sources, the debate of whether the telecommunication towers or RFEMF posed any health hazard or not has never ceased to disappear from the mainstream media or among scientists and researchers. Although there are published studies on electromagnetic fields in Malaysia, their focus was mostly on the fields of natural or physical sciences. For example, Ismail et al. (2009) documented that the readings of electromagnetic exposure at two major cities in Malaysia were well-below the guidelines set by the International Commission on Non-ionizing Radiation Protection (ICNIRP). Similarly, Pradhan et al. (2012) reported theoretical issues on electromagnetic field in Malaysia.

Therefore, this study was different because instead of natural or physical sciences, the issue of electromagnetic field was being examined from social sciences perspective. Essentially, this study seeks to investigate the public's perception of RFEMF effect on their health and well-being. Understanding the impact of perceived risk on well-being is essential in order to investigate the explanatory value and effectiveness of interventions influencing these beliefs. Therefore, this research examined the reliability and predictive validity of different risk perception operationalization related to RFEMF.

A combination of qualitative and quantitative approaches has been employed in the study. The mixed method approach is used simultaneously for the purpose of data collection and analysis, which consists of Phase 1 (qualitative approaches) and Phase 2 (quantitative approaches). Qualitative and quantitative data collection took place from January 2012 to August 2013. At the closing of the data collection process, a total of 2,245 samples were collected. However, after the data cleaning process, only 1,975 samples qualified for further analysis.

1.4.2   Research Scope

The study was conducted through the distribution of self-administered questionnaires based on the theoretical framework of Social Amplification of Risk (SARS) developed by Kasperson et al. (1988). Respondents who were participating in the study were asked about their perception with regards to electromagnetic fields. It means that the scope of questionnaires was not limited to questions regarding telecommunication towers only, as the study was also interested in gaining public perception and understanding of RFEMF.

The samples for the study were the general public. Therefore, respondents for the study were residents of all the states in Malaysia. It means that the sampling was comprehensive in terms of geographical coverage. Besides questionnaires, the study also included observations, face-to-face interviews and group discussions with other stakeholders such as the Government bodies, telecommunication service providers as well as consumers. It means that the study was comprehensive in terms of stakeholders' involvement.

1.4.3 Research Aims

Drawing upon the Social Amplification of Risk Framework (SARF), this study investigates the perception of RFEMF health risks by the public. SARF is the most comprehensive tool available for the study of risk. Theories and frameworks are useful and effective only to the extent that they conform to certain fundamental features of logic: clearly defined terms, coherence, internal consistency, sound organisation explicated with parsimony, accompanied by a specification of scope conditions and the generation of testable hypotheses.

Some people perceive risks from RFEMF exposure as likely and possibly even severe. Several reasons for public fear include social media interactions, word of mouth and media announcements of new and unconfirmed scientific studies, which lead to a feeling of uncertainty and a perception that there may be unknown or undiscovered hazards. Other factors are aesthetic concerns and a feeling of lack of control or input to the process of determining the location of new base stations.

Risk perceptions play a pivotal role in health behaviour theories. Understanding the impact of perceived risk on well-being is essential in order to investigate the explanatory value and effectiveness of interventions influencing these beliefs. This study investigated the reliability and predictive validity of different risk perception operationalization related to RFEMF in order to explain the inconsistent findings in the literature regarding the relationship between risk perceptions and public behaviours.

The relationship between risk knowledge, attitude and behaviour among different population groups is complex and has not been sufficiently explored (Khachkalyan 2006). It was essential to investigate this relationship in order to predict future behaviour for the well-being of the population. A better understanding of the influence of this knowledge on attitudes and risk reduction in a particular culture or country could significantly help in designing effective educational and awareness programmes and health messages. Essentially, this study of risk perception and its effect on the well-being of Malaysians was designed to fill the research gap and create a basis for further research in this area.

References

Akdag, M. Z., Dasdag, S., Canturk, F., Karabulut, D., Caner, Y., & Adalier, N. (2016). Does prolonged radiofrequency radiation emitted from Wi-Fi devices induce DNA damage in various tissues of rats? *Journal of Chemical Neuroanatomy, 75*(Pt B), 116–122.

Alhakami, A. S., & Slovic, P. (1994). A psychological study of the inverse relationship between perceived risk and perceived benefit. *Risk Analysis, 14,* 1085–1096.

Brown, V. J. (2014). Risk perception: It's personal. *Environmental Health Perspectives, 122*(10), A276–A279.

Cember, H. & Johnson, T. E. (2009). *Introduction to health physics* (4th ed.). McGraw-Hill Medical.

Covello, V., & Sandman, P. (2001). Risk communication: Evolution and revolution. In A. Wolbarst (Ed.), *Solutions to an environment in peril.* Baltimore: Johns Hopkins University Press.

Davis, F. S. (1989). Perceived usefulness, perceived ease of use, and user acceptance of information technology. *MIS Quarterly, 13,* 319–340.

Dohle, S., Keller, C., & Siegrist, M. (2012). Mobile communication in the public mind: Insights from free associations related to mobile phone base stations. *Human and Ecological Risk Assessment, 18,* 649–668.

EFHRAN. (2012). Risk analysis of human exposure to electromagnetic fields (revised). European Health Risk Assessment Network on Electromagnetic Fields Exposure.

Forbes, N., & Mahon, B. (2014). *Faraday, Maxwell, and the electromagnetic field: How two men revolutionized physics.* New York: Prometheus Books.

Geller, E. S. (2005). Behavior-based safety and occupational risk management. *Behavior Modification, 29,* 539–561.

Gorman, T., Dropkin, J., Kamen, J., Nimbalkar, S., Zuckerman, N., Lowe, T., et al. (2013). Controlling health hazards to hospital workers: A reference guide. *New Solutions, 23*(Suppl), 1–167.

Gray, P. (1998). Improving EMF risk communication and management: The need for analysis and deliberation, EMF risk perception and communication. In *Proceedings International Seminar on EMF Risk Perception and Communication, Ottawa, Ontario, CAN* (pp. 51–68).

Hardell, L. (2017). World Health Organization, radiofrequency radiation and health—A hard nut to crack (Review). *International Journal of Oncology, 51*(2), 405–413.

Heath, R. L., Liao, S.-H., & Douglas, W. (1995). Effects of perceived economic harms and benefits on issue involvement, use of information sources, and actions: A study in risk communication. *Journal of Public Relations Research, 7,* 89–109.

Hillson, D. (2004). *Effective opportunity management for projects—Exploiting positive risk.* New York, EE.UU: Marcel Dekker.

Ismail, A., Md Din, N., Jamaluddin, M. Z., & Balasubramaniam, N. (2009). Electromagnetic assessment for mobile phone base stations at major cities in Malaysia. In *IEEE 9th Malaysia International Conference on Communications.*

Jin, J.-M. (1999). *Electromagnetic analysis and design in magnetic resonance imaging.* Boca Raton, Florida: CRC Press.

Johnson, R. J., McCaul, K. D., & Klein, W. M. P. (2002). Risk involvement and risk perception among adolescents and young adults. *Journal of Behavioral Medicine, 25,* 67–82.

Kasperson, R. E., Ortwin Renn, P. S., Halina, B., Jacque, E., Robert, G., Jeanne, K., et al. (1988). The social amplification of risk: A conceptual framework. *Risk Analysis, 8*(2), 177–187.

Khachkalyan, T. (2006). Association between health risk knowledge and risk behavior among medical students and residents in Yerevan. *Californian Journal of Health Promotion, 4*(2), 197–206.

Kim, S., & Lee, J. (2012). E-participation, transparency, and trust in local government. *Public Administration Review, 72,* 819–828.

Kimura, T., Takahashi, K., Suzuki, Y., Konishi, Y., Ota, Y., Mori, C., et al. (2008). The effect of high strength static magnetic fields and ionizing radiation on gene expression and DNA damage in Caenorhabditis elegans. *Bioelectromagnetics, 29*(8), 605–614.

Lundgren, R. E. & McMakin, A. H. (2018). *Risk communication: A handbook for communicating environmental, safety and health risks*. Hoboken, New Jersey: Wiley.

National Research Council (US) Committee on Risk Perception and Communication. (1989). *Improving risk communication*. Washington (DC): National Academies Press.

Perrin, A., & Souques, M. (2012). *Electromagnetic fields, environment and health*. France: Springer.

Peters, R. G., Covello, V. T., & McCallum, D. B. (1997). The determinants of trust and credibility in environmental risk communication: An empirical study. *Risk Analysis, 17*(1), 43–54.

Pidgeon, N., & Henwood, K. (2010). The social amplification of risk framework (SARF): Theory, critiques, and policy implications. In P. Bennett, K. Calman, S. Curtis, & D. Fischbacher-Smith (Eds.), *Risk communication and public health* (2nd ed.). Oxford, UK: Oxford University Press.

Pina, V., Torres, L., & Royo, S. (2007). Are ICTs improving transparency and accountability in the EU regional and local governments? An empirical study. *Public Administration, 85*, 449–472.

Pradhan, A., Amirhashchi, H., & Zainuddin, H. (2012). A new class of inhomogeneous cosmological models with electromagnetic field in normal gauge for Lyra's manifold. *International Journal of Theoretical Physics, 50*, 56–69.

Purcell, E. M. (2012). *Electricity and magnetism* (3rd ed.). Cambridge: Cambridge University Press.

Qiu, W., Rutherford, S., Chu, C., Mao, A., & Hou, X. (2016). Risk communication and public health. *Global Journal of Medical Public Health, 5*, 1–11.

Renn, O., & Levine, D. (1991). Credibility and trust in risk communication. In R. E. Kasperson & P. J. M. Stallen (Eds.), *Communicating risks to the public*. Dordrecht: Kluwer.

Reyna, V. F., & Farley, F. (2006). Risk and rationality in adolescent decision making: Implications for theory, practice, and public policy. *Psychological Science in the Public Interest, 7*(1), 1–44.

Sjöberg, L. (1987). Risk, power and rationality: Conclusions of a research project on risk generation and risk assessment in a societal perspective. In L. Sjöberg (Ed.) *Risk and society*. London: Allen & Unwin.

Sjöberg, L. (2000). Factors in risk perception. *Risk Analysis, 20*(1), 1–12.

Slovic, P. (1987). Perception of risk. *Science, 236*, 280–285.

Slovic, P. (2010). *The feeling of risk: New perspectives on risk perception*. Abingdon: Earthscan, Taylor & Francis Group.

Starkey, S. J. (2016). Inaccurate official assessment of radiofrequency safety by the Advisory Group on Non-ionising Radiation. *Reviews on Environmental Health, 31*(4), 493–503.

Torcal, M. (2014). The decline of political trust in Spain and Portugal: Economic performance or political responsiveness? *American Behavioral Scientist, 58*, 1542–1567.

Tversky, A., & Kahneman, D. (1974). Judgment under uncertainty: Heuristics and biases. *Science, 185*, 1124–1131.

Utusan Online. (2008). Geran penyelidikan kaji kesan menara pemancar punca kanser (Research grant to study the impact of transmission towers on cause of cancer), 8 May 2008. Retrieved from www.utusan.com.my.

Utusan Online. (2018). Pemancar tak ancam kesihatan (The tower do no harm to health), 3 September 2018. Retrieved from www.utusan.com.my.

Wiedemann, P. M. (1999). *EMF risk communication: Themes, challenges and potential remedies* (pp. 69–94). Geneva: EMF risk perception and communication. WHO.

Wiedemann, P. M., Freudenstein, F., Böhmert, C., Wiart, J., & Croft, R. J. (2017). RF EMF risk perception revisited: Is the focus on concern sufficient for risk perception studies? *International Journal of Environmental Research and Public Health, 14*(6), 620.

World Health Organization. (2002). Radiation, Environmental Health, & World Health Organization. *Establishing a dialogue on risks from electromagnetic fields*. World Health Organization

World Health Organization. (2014). *Electromagnetic fields and public health: Mobile phones*. Retrieved from http://www.who.int/news-room/fact-sheets/detail/electromagnetic-fields-and-public-health-mobile-phones.

World Health Organization. (2015). *Risk communication saves lives & livelihoods: Pandemic influenza preparedness framework* [Brochure]. Author. Retrieved August, 2018, from http://www.who.int/risk-communication/PIP_brochure_EN_lo.pdf.

Yang, K., & Holzer, M. (2006). The performance-trust link: Implications for performance measurement. *Public Administration Review, 66,* 114–126.

Chapter 2
Design of the Study

Yusniza Kamarulzaman

2.1 Research Approaches

This study adopted qualitative as well as quantitative research design and subsequently integrated the findings from both approaches. The study was conducted for the duration of 24 months through two (2) phases. The first phase was a qualitative study through in-depth interviews and focus group discussions, while the second phase was a quantitative study via a questionnaire survey. This research combined the strengths of mixed-methods while enhancing the validity and reliability of the research findings. Figure 2.1 summarises the multi-methods approach carried out in this study.

Figure 2.1 shows the research design and approaches that were used correspondingly to fulfil the objectives of the study. This study relies on data sources which were gathered by both qualitative and quantitative research methods. Two data sources, primary data and secondary data, were collected, examined and analysed. The primary data were collected by both qualitative and quantitative research methods. Qualitative research methods involved an in-depth interview, focus group discussion and content analysis of the secondary data. Qualitative exploratory research methods are crucial to uncovering the prevailing issues in the thoughts and opinions of the study sample. This guides researchers to gain precise categorisation of the explored issues in specific dimensions and items. The following sections elaborate both phases and approaches.

2.1.1 Phase 1: Qualitative Study

The main aim of applying the qualitative approach in the first phase is to comprehend the scenario concerning public perception of RFEMF risk while uncovering the underpinning issues of the problems. These issues lead to the development

© The Author(s), under exclusive license to Springer Nature Singapore Pte Ltd 2020   17
Y. Kamarulzaman et al., *Public Perceptions of Radiation Effects on Health Risks and Well-Being*, SpringerBriefs in Environment, Security, Development and Peace 33, https://doi.org/10.1007/978-981-32-9894-1_2

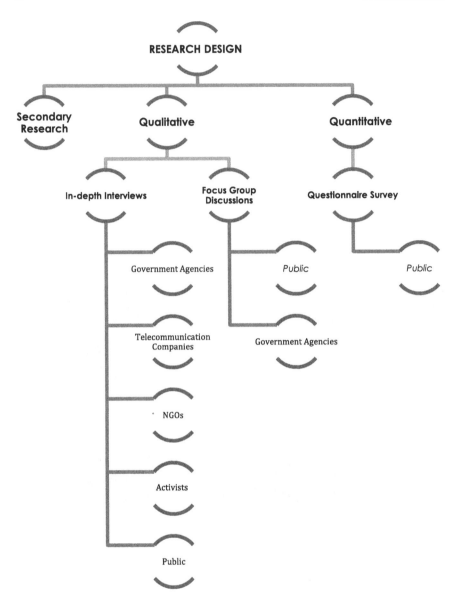

Fig. 2.1 Research Design and Approach

of research problems, questions and objectives which eventually act as the groundwork of the research framework. This qualitative phase is crucial as it provides guidance to develop the research framework and identifying research variables and research measures for the next phase, the quantitative study.

The qualitative phase of the study explored the risk perception and wellbeing of Malaysians in terms of health concerns. It is essential to investigate the perception of risk among public due to its impact on the future behaviour of a population. The relationship between risk knowledge, attitude and behaviour among different population groups is complex and has not been adequately explored (Khachkalyan 2006). A better understanding of the influence of public knowledge on attitudes and risk perception in a particular culture or country could significantly help in designing effective health awareness programmes and campaigns. Through the qualitative approach, this study was able to understand the scenario by exploring factors that shaped the perception of RFEMF risk among the public.

2.1.2 Phase 2: Quantitative Study

Data gathered from in-depth interviews and focus group discussions from Phase 1 was analysed and used to design Phase 2, the quantitative study. The aim of the Phase 2 was to find answers to the research questions and objectives derived from the research problems. This phase also aimed at developing and proposing a framework for this study and then subsequently testing the framework from the quantitative data gathered. All the variables and measurements were examined and tested in this phase to ensure the validity and reliability of the study. This phase was also meant to investigate any possible patterns and relationships across the variables being studied. Finally, this phase aimed at suggesting guidelines on policy and strategies related to RFEMF risk perception reduction effort for the authority to implement.

This quantitative study was conducted only among the public. A questionnaire survey was distributed face to face. Prior to data collection, the questionnaire was pre-tested and pilot tested to ensure the validity and reliability of the survey measurement. The research team travelled to the selected cities covering the north, south, west and eastern part of Malaysia including Sabah and Sarawak based on quota sampling. An online survey was also launched to complement the conventional survey in order to capture nationwide samples. The entire collected data was then inputted in SPSS for statistical analyses (Table 2.1).

Table 2.1 Respondents' profile for quantitative study

Demographics	Frequency	Percentage (%)
Gender		
Male	839	42.5
Female	1136	57.5
Ethnicity		
Malay	722	36.6
Chinese	738	37.4
Indian	290	14.7
Others	225	11.4

2.2   Research Samples

Figure 2.2 presents the sampling techniques and sample size of the study for both phases. Various types of samples were recruited for this qualitative study. The samples were from Government Agencies, Telecommunication Companies, NGOs and Activists and Publics. The samples were chosen based on their level of knowledge, experience and their roles related to RFEMF. Prior to the data collection process, each of the samples was given a consent letter.

A total of two focus group discussions with 5 and 6 participants respectively were conducted in two cities in the Northern and Southern of Malaysia where hot news about RFEMF appeared in the mass media. The participants comprised of the public and government agencies. Besides the focus group discussions, one-to-one in-depth interviews were also conducted across all sample types. The total number of samples of the in-depth interviews was 20 as summarised in Fig. 2.2.

The sampling method used for the quantitative study was cluster combined with random sampling which involved all states in Malaysia based on the affected areas and cases received by Malaysian Nuclear Agency (MNA). The sample size of each cluster was based on the number of cases reported in secondary data based on the number of development of telecommunication towers in Malaysia from 2010 to 2014. Subsequently, the sample size was established for each cluster based on the ethnic group population of the selected cluster. The rationale of choosing this technique was to ensure that the respondents had some knowledge and

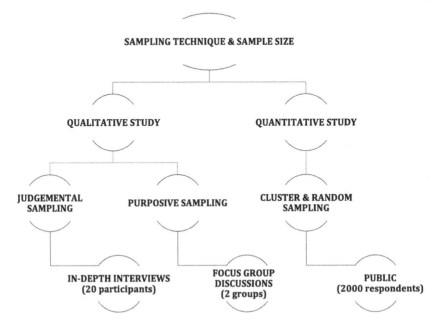

Fig. 2.2 Sampling Techniques

Table 2.2 Respondents by regions

Region	Target sample size	Actual sample size	Response rate (%)
Northern region	429	348[a]	79
East coast region	464	208	
Central region	822	922[a]	
Southern region	178	286[a]	
East Malaysia	607	186	
Not specified	–	25	
Total	2,500	1,975	

[a]*including the online survey*

understanding of the RFEMF issues, thus ensuring the generalisation of the findings across all states of Malaysia. Given a fixed budget, this would allow an increased sample size if needed. Assuming a fixed sample size, the mixture sampling methods led to highly representative samples that would better explain the variation in the population. Tables 2.2 and 2.3 show the distribution of respondents according to regions and states in Malaysia. Target sample size shown in Table 2.2 represent the number of the questionnaire survey distributed across the selected region. The study targeted 2,500 samples which was conducted conventionally. At the same time, the study also launched online survey via snowballing technique. Online survey respondents mainly came from Northern, Central and Southern Regions. The online survey responses were included in the actual sample size as presented in Table 2.2. From both surveys, a total of 1,975 responded to the survey. The distribution of respondents was considered fairly representative where the Northern, Central and

Table 2.3 Respondents by states and regions

Regions	States	Target sample size	Actual sample size
Central region	Klang valley	822	922
Northern region	Kedah	36	80
	Penang	143	105
	Perak	214	159
	Perlis	36	4
Southern region	Melaka	71	61
	N. Sembilan	36	91
	Johor	71	134
East coast region	Kelantan	107	78
	Pahang	286	124
	Terengganu	71	6
East Malaysia	Sabah	143	43
	Sarawak	464	143
Not specified	Not specific	–	25
Total		2,500	1,975

Southern regions contributed more than 50% of the total targets' sample size. However, there was low representation from East Coast and East Malaysia. Overall, response rate achieving almost 80% of the target sample throughout Malaysia was considered a good result.

Table 2.3 explains further by showing the sample size from each state of Malaysia. Being the capital of Malaysia, Kuala Lumpur, within Klang Valley has the biggest number of population. This is the reason the study received the highest number of respondents from Klang Valley which was 922 followed by Perak, Sarawak, Johor, Pahang and Penang.

2.3 Data Sources

2.3.1 Secondary Data

Through content analysis, the secondary data were extracted from previous research reports including journals, internet, newspapers, textbooks as well as documents produced by relevant organisations. It began with the exploration of reports related to RFEMF from the authorities such as the World Health Organisation, U.S. Federal Communications Commission, Malaysian Communications and Multimedia Commission (MCMC), Malaysian Nuclear Agency (MNA) as well as the media. Afterwards, related research papers were examined and pertinent issues related to risk perception of RFEMF were extracted.

2.3.2 In-depth Interviews

The in-depth interviews were carried out to explore insights into the main issues pertaining to risk of RFEMF. The interview questions were developed based on literature and established frameworks to ensure that different aspects of risk perception were covered. The interview participants were chosen from the Government Agencies, Telecommunication Companies, NGOs, activists and public.

The interview questions were developed based on literature and established models to include different aspects of risk perception. Questions were derived from SARF framework (see Chap. 1) as a guide for exploring the issue of RFEMF and risk perception. Each interview lasted for around 50 min to 1 h depending on the situations and type of participants. As mentioned above, 20 interviews were conducted amongst the government agencies, industry and the public (see Fig. 2.1). All team members were involved as interviewers while research assistants (RA) facilitated the team in organising the interviews, correspondences, telephone interviews, transcriptions and data management throughout Phase 1. The results of the in-depth interviews were the basis for the development of quantitative study which was used in the subsequent part of the research which is a focus group discussion.

2.3.3   Focus Group Discussions

The second method of data collection in Phase 1 of the research was focus group discussion (FGD). The advantages of focus group discussion over in-depth inter-view are the ability of the participants to build on or develop each other's argu-ments in order to bring out a deeper understanding and insights with regards to any point brought up by the participants during the discussion. The information col-lected from FGD was richer and in more concrete form as compared to the infor-mation gathered from individual in-depth interviews. The validity of the arguments was cross-confirmed among the participants themselves.

A total of two FGDs were carried out on two main groups (see Table 3.1). These groups are the participants from the public and government agencies. FGDs were carried out in Kuala Lumpur for the government agency and Ipoh for the public group. In the FGD sessions, the moderators first conducted an ice-breaking session with the participants followed by posting general questions about RFEMF and their understanding. The issues were gradually explored based on the answers and opinion shared by the participants. The FGD sessions were audio recorded with prior consent from the participants. Small tokens of appreciation were given to all participants.

2.3.4   Questionnaire Survey

The assumption in the study was that the general public was already aware of the risk that might be associated with the RFEMF. Therefore, the questions asked in the survey were not about whether people are aware of the risk of RFEMF; rather the questions were about the perceived degree of risk and the reasons why the risk can be associated with the RFEMF.

The questionnaires were distributed to the general public randomly through cluster sampling based on geographic locations. This will ensure that the ques-tionnaire will receive a good response. Therefore, the sampling of the general public was a mixture of simple random and cluster sampling. Besides personal contacts, questionnaires were also distributed to the visitors at one of the exhibitions held in Kuala Lumpur city.

This survey instrument was developed and continuously revised over time to include comments and results from the pilot-test and pre-test of the survey. The survey distribution was made through online and conventional (offline) methods. The survey was distributed across Malaysia including Sabah and Sarawak for more representative samples. A total of 2,000 samples were targeted with 1,000 samples

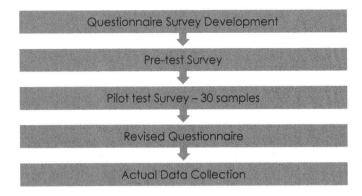

Fig. 2.3 Survey Stages

from offline or printed survey and another 1,000 samples were through an online survey. The surveys were distributed among Malaysian universities staff and students, activists and environmentalists, residents of selected areas throughout Malaysia, school teachers, parents, hospitals, organisations and the general public.

The online survey was disseminated through the project website (https://sites. google.com/a/um.edu.my/rfemf/), Facebook page and e-mail invitation from various sources based on a snowballing approach. Figure 2.3 presents a summary of the survey development schedule. Table 2.4 summarises the overall data source. The findings of the survey would be presented in Chap. 3.

Table 2.4 Summary of data source

Method	Details	Data source
Secondary data	Content analysis of data extracted from previous research reports related to RFEMF	Journals, internet, newspapers, textbooks as well as documents produced by relevant organisations
		Taken from authorities: World Health Organisation (WHO), U.S. Federal Communications Commission, Malaysian Communications and Multimedia Commission (MCMC), Malaysian Nuclear Agency (MNA) and the media
In-depth interviews	Questions were developed based on literature and established frameworks	Participants from the government agencies, telecommunication companies, NGOs, activists and the public
	20 interviews conducted, each took between 50 and 60 min	

(continued)

Table 2.4 (continued)

Method	Details	Data source
Focus group discussions (FGDs)	The information collected was richer and in more concrete form as compared to those gathered from individual in-depth interviews. The validity of the arguments was cross-confirmed among the participants themselves	Participants from a government agency and a public group
	Two FGDs were carried out on two main groups	
Questionnaire survey	The questions were about the perceived degree of risk and the reasons why the risk can be associated with the RFEMF	Participants were the general public, visitors at one exhibition in KL city

2.4   Data Analysis Techniques

2.4.1   Qualitative Analysis

The qualitative study was analysed using NVivo qualitative software. Interviews and focus group discussions were transcribed in verbatim and uploaded to the software for analysis. Themes were created based on the feedback (i.e. phrases, words, quotes, stories, body language etc.) from the participants. The themes were further tested and validated before the generalisation of the results was made.

Based on the review of secondary data and qualitative data analysis, a research framework was established to shape the direction of the study as shown in Fig. 2.4. This preliminary research framework presents the outline of the study which encompasses predictors that influence the perceived risk of RFEMF and eventually the behavioural outcome.

Generally, the findings of the qualitative study show that public perception of RFEMF risk is influenced by factors such as:

- Emotional responses (anger and fear)
- Media influence (media content, availability, usage, word of mouth)
- Behavioural belief and control
- Socially responsible consumption
- Health conscious
- Attitude towards development
- Material values
- Subjective norms
- Political influence.

These factors will be further discussed in Chap. 3. These findings were used as the basis of the quantitative study framework as shown in Fig. 2.4. These factors

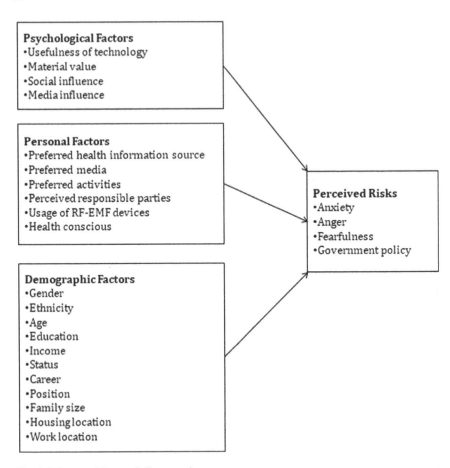

Fig. 2.4 Proposed Research Framework

were further investigated in larger sample size (1,975 samples) through survey questionnaires.

2.4.2 Descriptive Analysis

To better understand the characteristics of the sample, descriptive statistics analysis was applied to illustrate the respondents' characteristics, means and standard deviation of each variable. The descriptive analysis was also used to identify the profile of the respondents and to answer part of the research questions.

2.4.3   Reliability and Validity Test

Reliability and validity are of vital importance in the measurement scales. Reliability analysis is a measure of the internal consistency of indicators for a construct (Hair et al. 2009). The purpose of reliability analysis is to determine how well a set of items taps into some common sources of variance and is frequently measured with Cronbach's coefficient alpha. Cronbach's coefficient alpha is "the ratio of the sum of the covariances among the components of the linear combination (items), which estimates true variance to the sum of all elements in the variance-covariance matrix of measures, which equals the observed variance" (Nunnally and Bernstein 1994). Validity is the extent to which a scale or set of measures accurately represents the concept of interest. Content validity was established during preparation of the questionnaire by means of scales already validated in the literature (Devellis 2003) matched with the findings from Phase 1.

2.4.4   Correlation and Multiple Regression Analysis

Correlation analysis was used to investigate the relationships between the independent and dependent variables. This test was conducted to identify the significance, the strength and the magnitude of each relationship. Afterwards, multiple regression analysis was conducted to test the explanatory power of the proposed model of RFEMF risk perception. The regression analysis determined the factors that influence the RFEMF risk perception the most, the best predictor of RFEMF perceived risk model.

2.5   Reliability and Validity Analysis

Besides the demographic variables as shown in the earlier section, there were 123 variables with regards to the measurement of perceived risk of RFEMF. In order to ensure the quality of the primary data that we have collected, rigorous data screening, reliability and validity tests were performed. Data screening was conducted to ensure the data was clean through descriptive statistics, normality test, and analysis of outliers. The reliability of data was represented by Cronbach's alpha. Its value ranges from 0 to 1 and a value of more than 0.60 was considered highly reliable. It was reliable in the sense that the scales used in the survey have measured the constructs properly. If the Cronbach's alpha was low, it meant that the reliability of the construct was low. In order to increase the reliability of the constructs, some scales have been removed or some constructs may have been merged.

Table 2.5 shows that all the 18 Cronbach's alphas are greater than 0.60, so all scales are relatively reliable and can be used to measure the construct. For example,

Table 2.5 Reliability and factor analysis

Construct	Item	Item loading	Cronbach's Alpha
Perceived risk: government policy related	P3. Bad versus Good	0.616	0.934
	P4. Worthless versus Valuable	0.587	
	P1. Unsatisfactory versus Satisfactory	0.566	
	P7. Bias versus Fair	0.526	
	P2. Static versus Growth	0.516	
	P5. Declining versus Improving	0.490	
	P6. Restriction versus Flexibility	0.443	
Perceived risk: fearfulness	F3. I feel afraid when I think of RFEMF radiation risk	0.432	0.891
	F1. When I think of RFEMF radiation risk, I have a fearful gut feeling	0.427	
	F2. When I see an RFEMF radiation risk, I become scared	0.400	
Perceived risk: anger	AG3. When I see telecommunication development, I get angry easily	0.424	0.820
	AG2. I simply get irritated in a discussion about telecommunication development with other people	0.397	
	AG1. I feel angry when a new mobile phone base station will be constructed	0.316	
Perceived risk: anxiety	U2. Sometimes I feel restless or stressed out about problems associated with RFEMF radiation risk	0.460	0.898
	T1. I often feel anxious about RFEMF radiation risk	0.460	
	T2. I sometimes feel physical symptoms of nervousness / anxiety when I think about RFEMF radiation risk	0.418	
	U4. I get over-excited when I talk about RFEMF radiation risk	0.415	
	T4. I worry a lot about problems associated with RFEMF radiation risk	0.399	
	T3. Sometimes I get so worried about RFEMF radiation risk and I can't get it out of my head	0.386	
	U3. When I think about RFEMF radiation risk, I will feel tense	0.359	
	U1. Discussing / Discussing/thinking radiation risk makes me irritable	0.343	
Behavioural outcome: intention to continue using the devices	IT5. I intend to accept telecommunication development in future	0.569	0.851
	IT1. I wanted to support telecommunication development rather than being against it due to its benefits	0.491	
	IT2. My choice is towards telecommunication development than maintaining the current status	0.447	
	IT3. For managing my personal matters, I intended to use telecommunication as much as possible	0.396	
	IT4. I expect to support telecommunication development in every project that I know	0.394	

(continued)

Table 2.5 (continued)

Construct	Item	Item loading	Cronbach's Alpha
Behavioural outcome: attitude towards RFEMF technology	AT4. Worthless versus valuable	0.599	0.900
	AT3. Bad versus Good	0.597	
	AT2. Unpleasant versus pleasant	0.571	
	AT1. Harmful versus beneficial	0.567	
	AT5. Un-Enjoyable versus Enjoyable	0.457	
Behavioural outcome: self-control	BC2. For me to protest the telecommunication development is my right	0.408	0.696
	BC5. I had the resources, knowledge and ability to use telecommunication facilities	0.381	
	BC1. Whether I accept / reject the telecommunication development is entirely up to me	0.372	
	BC3. I would be able to use telecommunication facilities for managing my life	0.357	
	BB2. It causes worries / concerns amongst the society if they found telecommunication tower built nearby their residential area	0.326	
	BB1. If I protest against the telecommunication development, I will feel that I am doing something positive for the society	0.318	
	BC4. Using telecommunication facilities was entirely within my control	0.313	
Behavioural outcome: socially responsible action	B1. People should be more concerned about reducing the RFEMF radiation risk	0.497	0.812
	B2. Every person should stop increasing his / her consumption related to RFEMF radiation	0.473	
	B3. RFEMF radiation risk is presently one of the most critical problems	0.466	
	B10. Schools and colleges should require all students to learn about RFEMF radiation risk	0.466	
	B7. Commercial advertising should be forced to mention the RFEMF radiation risk	0.449	
	B8. People should encourage their friends to avoid RFEMF radiation risk	0.448	
	B5. I would be willing to sign petition / demonstrate to save the environment from RFEMF radiation risk	0.442	
	B4. RFEMF is not personally affecting my life	0.338	
	B9. I will stop buying products related to RFEMF radiation risk even it might be inconvenient	0.298	
	B6. I'd be willing not to use mobile phones in order to reduce RFEMF radiation risk	0.285	
Psychographic factors: material value	M4. It's really true that money can buy happiness	0.407	0.806
	M2. I would like to be rich enough to buy anything I want	0.388	
	M3. I'd be happier if I could afford to buy more things	0.319	
	M1. It is important to me to have really nice things	0.311	

(continued)

Table 2.5 (continued)

Construct	Item	Item loading	Cronbach's Alpha
Psychographic factors: usefulness	US3. Telecommunication development would enhance the country's effectiveness	0.448	0.873
	US2. Telecommunication development would improve the productivity of the community	0.441	
	US1. Telecommunication development would improve the overall performance of the nation	0.419	
Psychographic factors: social influence	SN1. People (peers and families) supported my views on RFEMF risk	0.447	0.649
	WM3. I speak of telecommunication development's good sides	0.383	
	SN2. People who influenced my behaviour wanted me to support telecommunication development instead of being against it	0.343	
	WM1. I spoke of RFEMF radiation risk to many individuals	0.340	
	WM2. I recommended telecommunication development	0.335	
	SN3. Most people who are important to me think that telecommunication development is risky	0.309	
Psychographic factors: media influence	ME3. Risks of RFEMF radiation are reported fairly in the media	0.385	0.683
	EI3. Mass media reports influenced me to try out new telecommunication facilities	0.345	
	EI2. The popular press depicted a positive sentiment for telecommunication development	0.342	
	EI1. I read/saw news reports that telecommunication development was good for the nation	0.341	
	ME1. Media exaggerate the risks of RFEMF radiation in their reports	0.323	
	ME2. Media ignore the risks of RFEMF radiation in their reports	0.312	
Personal: preferred health information sources	MS13. Events / Campaigns / Exhibitions	0.638	0.893
	MS11. Interest Groups (e.g.:- Environmental groups, Activists)	0.584	
	MS3. Social Media (e.g.:- Blogs, Facebook, Twitter)	0.555	
	MS10. Other Consumers / Public	0.507	
	MS12. Family / Friends	0.489	
	MS1. Internet websites	0.462	
	MS9. Government agencies	0.459	
	MS8. Radio	0.450	
	MS7. Television	0.417	
	MS5. Newspapers	0.404	
	MS14. Others	0.401	
	MS6. Magazine	0.395	
	MS4. Mobile Phone (e.g.:- SMS, MMS)	0.362	
	MS2. Your Doctor	0.325	

(continued)

Table 2.5 (continued)

Construct	Item	Item loading	Cronbach's Alpha
Personal factors: health conscious	H4. I am aware of the state of my health every day	0.409	0.867
	H1. I am self-conscious about my health	0.375	
	H3. I am usually aware of my health	0.366	
	H5. I am involved with my health	0.349	
	H2. I am constantly examining my health	0.334	
Personal: preferred responsible parties	G9. Local community groups	0.507	0.913
	G6. National and state	0.485	
	G8. Private industry	0.484	
	G7. Environmental groups	0.471	
	G4. Local council	0.460	
	G5. Individual citizens	0.421	
	G10. Others	0.399	
	G2. Government departments of the environment	0.390	
	G3. Medical Doctors	0.382	
	G1. Government departments of health	0.345	
Personal: preferred activities	AC3. Socialising (e.g.:- attend gathering, reunion, association, informal meetings etc…)	0.470	0.771
	AC7. Sports and games (e.g.:- gym, fitness class, indoor, outdoor etc…)	0.460	
	AC4. Knowledge seeking (e.g.:- attend training workshops, seminars, lectures etc…)	0.442	
	AC5. Social service work (e.g.:- orphanage, hospital, school, CSR etc…)	0.437	
	AC1. Internet browsing and interaction (e.g.:- email, chatting, Facebook etc…)	0.393	
	AC6. Reading (e.g.:- books, newspaper, magazines etc…)	0.371	
	AC8. Entertainment (e.g.:- music, TV, cinema, karaoke etc…)	0.357	
	AC2. Telephone interaction (e.g.:- phone calls, video calls, SMS, MMS etc…)	0.332	
Personal: preferred media	PM5. Social Media (e.g.:- Facebook, Twitter etc…)	0.477	0.643
	PM1. Newspaper (e.g.:- Berita Harian, The Star etc…)	0.385	
	PM2. Magazine (e.g.:- Remaja, Cleo, Wanita etc…)	0.381	
	PM4. Radio (e.g.:- Sinar, Hitz.fm, BFM etc…)	0.367	
	PM3. TV channel (e.g.:- ASTRO, TV3, National Geographic etc…)	0.351	

(continued)

Table 2.5 (continued)

Construct	Item	Item loading	Cronbach's Alpha
Personal: Usage of RFEMF Devices	US14. Wireless devices (e.g. modem, mouse, alarm, printer, baby monitor etc.)	0.448	0.816
	US13. Smart metre (measures energy consumption)	0.446	
	US6. Personal computer / Laptop	0.414	
	US3. Radio	0.408	
	US4. Television	0.382	
	US10. Bluetooth devices	0.368	
	US9. Wi-Fi routers	0.359	
	US11. Video game consoles	0.357	
	US1. Cell Phone / Smartphone	0.348	
	US2. Cordless Phone	0.340	
	US5. Microwave oven	0.304	
	US12. Medical / laboratory equipment	0.299	
	US7. iPad / Tablet / PDA	0.290	
	US8. Digital alarm clock	0.266	

"Perceived Risk: Government Policy Related Initially" construct was measured by seven items (questions) and its Cronbach's alpha was 0.934. It was an indication that the questions were reliable in measuring the construct. However, Cronbach's alpha for "behavioural belief" was initially 0.506 and this was considered low. It has been merged with "behavioural control" and was called "Self-Control". The new construct Cronbach's Alpha was 0.696 and was considered more reliable.

Concurrent with the reliability analysis, we also performed factor analysis. The purpose of the factor analysis was basically to reduce the number of determinants so that the reduced determinants could explain the variation in the unobserved variable. We examined three aspects of factor analysis, which were variable loading, Kaiser-Meyer-Olkin (KMO) and Bartlett's test.

The value of variable loading ranges from -1 to 1, where 1 indicates strong positive relationship, while -1 means strong negative relationship. A variable loading greater than 0.30 meant that the factor exhibited a strong and good relationship. Table 2.5 clearly shows that only 5 items had factors below 0.3. However, when we compared the loading with the reliability test, Cronbach's Alphas of related constructs were mostly reliable. Hence, we decided to include these 5 items in the analysis.

The value of KMO also ranges from 0 to 1. Similarly, a high value of KMO meant that the scales were adequate in explaining the factors in question; while a low value of KMO meant that the scales cannot adequately explain the factors. The value of KMO is 0.813, thus the scales were deemed to be adequate and there was no necessity to drop any variables.

In addition, from the factors results, it was recommended that our study should have 20 constructs as it explains 60% of the total variance. It aligned with our existing framework. However, by looking at the Scree Plot below, 16 constructs explained 55% of the total variance. Hence, due to the low Cronbach's Alpha for the "behavioural belief" construct, it was merged with "behavioural control" to form the "self-control" construct. As in the results, finally we had 18 constructs (within the recommended range of SPSS) for this study and proceeded for further analysis.

Another indicator of the strength of the relationship among variables was Bartlett's test of sphericity. Bartlett's test of sphericity was used to test the null hypothesis that the variables are uncorrelated. The observed significance level was 0.00, so it was small enough to reject the hypothesis. It meant that the strength of the relationship among variables was strong. Therefore, we proceeded with factor analysis.

As mentioned earlier, there were more than 100 variables in the study. The purpose of conducting factor analysis was to reduce the number of variables so that the unobserved variable could be explained; so, the variables were reduced to 18 as shown in Table 2.6. Personal factors, psychological factors, perceived risk and behavioural outcome were four unobserved variables (factors) that could be derived from the constructs. For example, "Perceived Responsible Parties" was one of the constructs that formed "Personal factors".

With the minimum score of 1 and a maximum of 5, the mean scores for each construct were relatively high, more than 3, except for two constructs. However, all constructs had high reliability. Therefore, two constructs were still acceptable. In

Table 2.6 Reliability test of constructs

Measurements	Mean	Std. Dev.	N (item)	Cronbach's Alpha
Personal factors				
1. Perceived responsible parties	4.02	6.45	10	0.913
2. Health consciousness	3.90	3.33	5	0.867
3. Preferred media	3.45	3.57	5	0.643
4. Preferred activities	3.38	5.30	8	0.771
5. Preferred health information source	3.21	10.27	14	0.893
6. Usage of RFEMF devices	2.95	9.65	14	0.816
Psychological factors				
1. Usefulness of technology	3.93	2.03	3	0.873
2. Material value	3.36	3.48	4	0.806
3. Media influence	3.36	3.25	6	0.683
4. Social influence	3.19	3.13	6	0.649

(continued)

Table 2.6 (continued)

Measurements	Mean	Std. Dev.	N (item)	Cronbach's Alpha
Perceived risk				
1. Government policy	3.28	5.85	7	0.934
2. Fearfulness	3.25	2.69	3	0.891
3. Anxiety	3.05	6.19	8	0.898
4. Anger	2.90	2.14	3	0.820
Behavioural outcome				
1. Attitude Towards Telecommunication Development	3.28	5.85	7	0.934
2. Self-Control	3.25	2.69	3	0.891
3. Intention to Continue Using the Devices	3.05	6.19	8	0.898
4. Socially Responsible Action	2.90	2.14	3	0.820

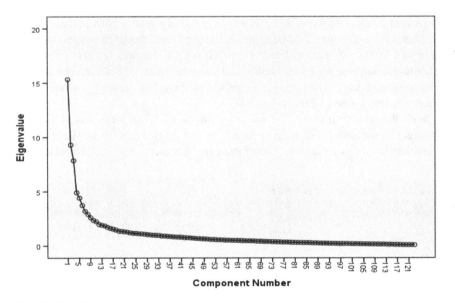

Fig. 2.5 Scree Plot

addition to the results in Tables 2.5 and 2.6, the scree plot in Figure 2.5 shows that the recommended constructs should be between 13 and 21. For example, 20 constructs could explain 60% of the total variance, while 15 constructs could explain 55% of variance. In line with literature review we mentioned earlier, we finally accepted 18 constructs.

2.5.1 Regression Analysis

Regression analysis was used to test the best fitness of the model itself, i.e. how good the conceptual framework proposed influence perceived risks. Table 2.7 shows that the model is statistically significant with adjusted R^2 of 0.334. This meant that the model can explain 33.4% of the variance in perceived risks where all eight factors were statistically significant; namely, media influence, social influence, work category, ethnicity, preferred health information source, material values, usefulness of technology and health consciousness. All factors were significant at the 0.05 level.

It was noted that the coefficient of ethnicity was negative and since "1" referred to Malay ethnicity, it meant that the Malays perceived risk of RFEMF was higher than other ethnicities. Similarly, people who thought that the technology was not useful had a higher perceived risk of RFEMF than people who thought that the technology was useful. It was also noted that social influence had the highest influence among all eight factors. This showed that social context highly contributed to the perceived risk of RFEMF. This was followed by media that could highly influence people's perception of risk.

Factors in Table 2.7 were grouped under several factors. The preferred health information source and health conscious were grouped under "personal factors", while "psychological factors" included social influence, media influence, material values and usefulness of technology. Finally, "demographic factors" involved the work category and ethnicity.

Table 2.7 The results of regression model of perceived risks (n = 1,975)

Factor	Unstandardized coefficient, B	t-value
(Constant)	1.412	12.879**
Media influence	0.167	6.570**
Social influence	0.428	16.374**
Work category	−0.006	−2.220
Ethnicity	−0.030	−2.375
Preferred health information source	0.113	7.116**
Material values	−0.031	−2.459
Usefulness of technology	−0.202	−11.791**
Health consciousness	0.106	6.468**

Dependent variable: perceived risk

**significant at p < 0.001

2.6 Multivariate Analysis of Variance

Multivariate analysis of variance (MANOVA) was an extension of analysis of variance when we had more than one dependent variable. MANOVA compared the groups and articulated us whether the mean differences between the groups on the combination of dependent variables were likely to have occurred by chance.

In this section, the perceived risk was divided into four dependent variables called anger, anxiety, fearfulness and government related. Basically, it meant that people's anger with regards to RFEMF risk could be used to indicate perceived risk. The purpose of MANOVA was to examine the impact of psychological factors, personal factors and demographic factors towards these four dependent variables.

References

Devellis, R. F. (2003). *Scale development: Theory and applications*. Thousand Oaks, CA: Sage Publications.

Hair, J. J. F., Bush, R. P. & Ortinau, D. J. (2009). *Marketing Research: In A Digital Information Environment* (4 edn.), McGraw-Hill International Edition.

Khachkalyan, T. (2006). Association between health risk knowledge and risk behavior among medical students and residents in Yerevan. *Californian Journal of Health Promotion, 4*(2), 197–206.

Nunnally, J. C., & Bernstein, I. H. (1994). *Psychometric theory* (3rd ed.). New York: McGraw-Hill.

Chapter 3
Qualitative Findings: Exploring RFEMF Risk Perceptions

F. D. Yusop

3.1 The Participants

Samples of the study are individual stakeholders who are anxious about the risks posed by the telecommunication towers and employees of government agencies and telecommunication companies who are concerned about people's reaction.

Apparently, we managed to locate a public forum about the risk of radiofrequency which was held in the country in the middle of 2012. Organised by the activists, we participated in the forum and were able to meet potential respondents who fulfilled the sample characteristics. In total, this study involved 25 individuals (Table 3.1); consisting of parents and teachers, including two employees of telecommunication companies and five employees of Government agencies. The inclusion of the employees will provide better understandings about the issue under study. We observed ten public schools, one public forum and meeting of government agencies.

The two RFEMF experts who were interviewed are well known figures in the area of RFEMF in the country and internationally. Both experts have academic backgrounds in the area of risk communication and actively consulted various government agencies in the area. As such, they have been involved in giving talks, producing guidelines and writing reports on the topic (Table 3.2).

We are well aware that some respondents are government servants, so they might not feel comfortable while expressing their opinion regarding a government project. However, we managed to assure them that the participants of this study remained anonymous unless they wish otherwise. All the interviews went well and respondents did not constrain themselves from giving their honest opinion. The interviews were then transcribed and the transcripts were coded in the way suggested by grounded theory approach (Glaser/Strauss 1967). Based on the audio data, respondents' concern was identified and its factors were generated. The factors were then cross-checked with data from our observations. Therefore, the entire data was accounted for in the analysis.

Y. Kamarulzaman et al., *Public Perceptions of Radiation Effects on Health Risks and Well-Being*, SpringerBriefs in Environment, Security, Development and Peace 33, https://doi.org/10.1007/978-981-32-9894-1_3

Table 3.1 List of interviews and focus group discussion from March to December 2012

No.	Interview/Respondents	No. of participants	Group/Sector
1	Government officers	2	Government agency
2	Government officers	4	Government agency
3	Commercial telecommunication staff	1	TelCo company
4	Telecommunication infrastructure development staffs	2	TelCo company
5	Regulatory, technical and media strategy	3	Focus groups
6	Doctor and activist	1	Doctor and activist
7	Activist	1	Doctor and activist
8	Commercial telecommunication staffs	2	Telecommunication company
9	Radio interview with RFEMF expert	1	Expert
10	Public (with a background of Information and Technology)	2	Public
11	Commercial telecommunication staff	1	Telecommunication company
12	Risk communication expert	1	Expert
13	Public	1	Public
14	Teachers	2	Teachers/Public
15	Teacher	1	School/Teachers

Table 3.2 Other sources of information

Type	Respondent(s)
Focus group discussion	Regulatory, technical and media strategy groups (4 respondents)
Presentations	Presentations by experts and environmental activists
Expert analysis	A radio interview with RFEMF expert
Document analysis	Analyses of government official letters, complaints etc. related to installation of telecommunication towers in school compounds

3.2 Concerns Towards RFEMF Risks

In this study, four main stakeholders were identified; firstly, the government agencies that regulate the telecommunication industry and have the authority in terms of enforcement of law; secondly, the telecommunication companies that provide the services to the consumers; thirdly, the non-government organisations (NGOs) and activists that are concerned about the risk of RFEMF, and finally the general public.

Based on the focus group discussions as well as face-to-face interviews with all relevant stakeholders, it is found that each group of stakeholders—the government agencies, telecommunication companies, the non-governmental organisations and the general public—have different perceptions and concerns towards RFEMF.

3.2.1   Concerns Among Government Agencies

The study finds that there is growing demand of the government agencies to examine the RFEMF radiation transmitted from mobile base station. This indicates an increase in public awareness and concerns regarding the EMF risk. Even though the Government agencies have limited human resources, current equipment and technologies, they are still able to fulfil the demand. At all the sites that they have examined, it is found that the RFEMF signal transmitted is very low, therefore following the ICNIRP guidelines is considered safe.

Even though readings of the RFEMF signals proved below the legal limit, the government agencies experience difficulties in convincing the public that the sites are safe. Most of the time, the public challenged their readings. It is believed that it is due to their pre-misconception of the risks which is highly influenced by rumours through word-of-mouth and social media. Additionally, the limited of formal and informal educational programmes by the government agencies to educate the public on RFEMF risks makes it difficult to refute the perceptions.

3.2.2   Concerns Among Telecommunication Companies

The study also interviewed staff of several telecommunication companies who have the decision-making power. The telecommunication companies staffs seem to be aware of public concerns but choose to ignore them as they believe that the risks associated with RFEMF are not severe. They also indicated that they rarely receive direct complaints from the public related to RFEMF risks and if they do, they will hire a governmental agency to measure the radiation level.

In their efforts to lower the cost and provide better telecommunication signals, telecommunication companies tend to share their towers with other telecommunication companies and install them in selected areas such as residential areas, business centres and schools. In the case of residential areas, there are some Muslim's prayer centres, or known as *surau* in local language, that have allowed towers to be built in their centers' compounds. In these cases, the telecommunication companies pay some fees to the *surau* committee who will then utilise the money to fund some of their *surau* activities. Since the installation of towers in the compounds was mutually agreed and mutually beneficial, there were almost no complaints made by the residents in the areas.

As the public demand for better telecommunication services continues to grow, the telecommunication companies project more towers to be built in the future to fulfil the public demand and needs. At the same time, they work closely with government agencies such as MCMC to help educate the public about the RFEMF risks as part of their corporate social responsibility (CSR) commitment.

3.2.3   Concerns Among Non-government Organisations (NGOs) and Activists

Contrary to the government agencies and telecommunication companies, NGOs and activist groups show a different perspective. The study found that they are actively disseminating information and educating the public on RFEMF radiation via seminars, activities, websites, columns etc. at grassroots level. These events were organised at national level which involved international and local speakers to create awareness among Malaysian public. Many of the NGOs members and activists are knowledgeable and experts in the area of RFEMF radiation. The NGOs also have strong connections with other international NGOs and RFEMF experts. Since they are closer to the public, they seem more trusted and their opinion is perceived as impartial. This is because they neither receive government funds nor have direct affiliations with the telecommunication companies.

The study finds that the NGOs and activists are not totally against installation of telecommunication towers as perceived by some government authorities and/or telecommunication companies. Their main concern is that the current standard for radiation exposure adopted in Malaysia is rather high—even if it does not exceed the ICNIRP standard—compared to other nations such as Austria. Additionally, the NGOs strongly advocate that the public should be made aware of any effort to install such towers near residential area and the general public should be consulted and involved in the decision-making process.

In short, the NGOs and activists are suggesting a moderate solution to the issue:

- reduce the radiation safety limit from 4.5 W/m^2 to the lowest level possible.
- avoid installing telco towers/gadgets in "sensitive areas" e.g. schools and residential areas.
- limit the installation of telecommunication towers to commercial areas only.
- focus on the 1/3 of the affected samples rather than making claims that it is harmless.
- invite the public in any decision related to installation of telecommunication towers.
- ask permission from the residents if some telecommunication company wants to build its tower near residential area.

The close connections of the NGOs and activists with the public make them powerful organisations representing the public voice. They too are willing to work together with the authorities and share their expertise in any policy-making activities and decisions regarding RFEMF. This makes the NGOs and activists highly trusted. Hence, we suggest that they should not be ignored and need to be involved in any discussions and activities with authorities in this matter.

3.2.4 Concerns Among the General Public

In contrary, the findings from the public show the diverse perception towards RFEMF risk due to knowledge about RFEMF among Malaysian public. People who are against the RFEMF are found to be well-informed and thoroughly did their own research. For example, some people claimed that the Malaysian government requirement sets the power density at 10 million microwatt per-meter square based on the recommendation of ICNIRP. This measurement, according to the activists, was based on the 1998s measurement which is now out-dated and no longer practiced in many other countries in the world. The local government of Salzburg, Austria, for instance, has cut down the numbers to as low as 100 $\mu W/m^2$. Therefore, this contrasting fact makes them question the government agencies so-called 'expertise' in RFEMF.

Some are doubtful about the government agencies' ability to manage the situation. Firstly, it is almost impossible for the government agencies to sense the radiation level when they have limited expertise, human power and updated tools to do so. Secondly, there is a question of whether the government agencies will take actions against the powerful and rich telecommunication companies.

3.3 Perceptions Toward RFEMF Risks

Depending on the knowledge people have about the risk, people perceive risk differently, hence contributing to the differences in perception of risks between the experts and the laypeople (Slovic 1987). Laypeople usually associate images they have in their mind with risks. For example, if they identify mobile base station with negative images, their perception of risks is high. On the other hand, if they have positive image associated with the mobile base station, their perceived risk is low (Dohle et al. 2012).

There are two forms of knowledge used by individuals to assess risk; (i) knowledge that people know; and (ii) knowledge that people think they know. Although people's perception of risk depends on their knowledge, people tend to use the latter as a guide of risk. Therefore, it is difficult to establish the relationship between knowledge and risk (Ferguson 2001). In this study, we enquired the basis of underlying reasons and found that people believe in certain sources of information which lead them in forming their perception on the risks.

3.3.1 Formation of Risk Perceptions

We found that people's judgments are closely related to the sources of information about the risks and whether or not the perceived risks outweigh its benefits.

(a) **Sources of information**

As the public can nowadays easily search and look for information, their perception is shaped depending on the sources of information they believed. Nonetheless, differences of judgments about risks between the public and the experts do occur which can be partly explained by biases in values and emotions. The experts always measure risks by the number of lives lost, while the public draws finer distinctions. For example, fatalities in random accidents are viewed differently from deaths that occur in the course of voluntary activities such as skiing or racing (Kahneman 2011: 140; Slovic 1987).

The study finds that the respondents are fearful of radiation based on the information they obtained from the Internet. Their perception is made as respondents are resourceful and always conscious about the radiation that may be linked to the telecommunication towers. A respondent had printed and showed us materials related to the strength of the signals and how it was measured and claimed that:

> I have seen and read more than 20 studies which show that if the distance is less than 200 meters, then the effect of radiation is eight times higher… I did not make this up… I can show you the studies (Parent).

Another respondent showed the standard operating procedures used for the installation of telecommunication towers in other countries. His perception has been formed based on the information he thought he knew and deemed credible. Besides, a respondent argued that there is an alternative technology such as fibre cable which is less costly and less risky because it can make use of the existing landline telephone network. Hence, the construction of telecommunication towers is not required for the project.

> Actually the schools do not need the telecommunication towers… you can use fibre optic cable and then there will be no radiation (Parent).

It is obvious that the respondents have made their own perception based on the information they highly trust although there are other reports that provide different opinions. It raises questions why people trust certain sources and do not believe in other sources of information. This is a challenge to the authorities because information provided by the governments is usually perceived less trustworthy (Frewer et al. 1996).

(b) **Perceived risks exceed benefits**

As suggested by Slovic (1987), our perceptions were influenced by our emotions when we rated the perceived benefits and risks of a technology. If we disliked it, we could only think of its disadvantages. For instance, our respondents agree that telecommunication technologies are important for 21st century education, however they opposed the technical aspects of telecommunication towers' installation that are too close to the classrooms, as they believe that the telecommunication towers will transmit radiofrequency radiation and are thus harmful to school community, especially children. As one of the teachers mentioned:

I have no objection to the project if we can minimise the risk ... the telecommunication towers are risky ... use fibre optic cable. (Teacher)

To our respondents, the benefits of the project are not sufficient enough to justify the risks that they have to bear. In this situation, the sources of information they received from many sources – credible or incredible – have formed their risk perception regardless of whether or not their perceptions are true and scientifically proven.

3.4   The Role of Trust in Shaping Risk Perceptions

In the society, trust is important to ensure that people can easily work with each other and create a sense of community. Trust is the attribute that bonds the relationship within members of the group or inter-groups. Trust can be broken and need to be earned as well. Hence, routines and practices are required to maintain trust (Solomon/Flores 2001: 24–26). In fact, when we entrust other people to perform a task for us, we actually underestimate the risk. Therefore, trust has become an important element in risk research because it is capable of shaping the perception (Uslaner 2007).

In a democratic state, the management of hazard is the state's responsibility. Although the authorities may think that they have done their job by disclosing relevant information to the public, it might not change perceptions especially when people are well-informed about the hazard (Siegrist/Cvetkovich 2000). Information from the industry and government often do not gain public trust because they are perceived less credible due to conflict of interests. Meanwhile, information obtained from independent sources, such as consumer organisation, is usually highly trusted (Frewer et al. 1996).

3.4.1   Development of Trust

Several factors influence public trust including effective communication, monitoring and control, project competency and media coverage.

(a) Communication about the project

Even though we find in the study that trust and credibility in the authorities are the public's main concern, it has become a chicken-and-egg problem in the sense that the respondents tend to trust the relevant authorities if they have good perception of the credibility of the authorities. However, if people perceive lack of credibility in the authorities, how can they trust the authorities? When two people do not trust each other, they will not talk to each other; therefore an honest communication between the two parties is an important virtue in trust building. Withholding critical information and shutting down communication do not help build trust (Solomon/ Flores 2001: 44).

In the study, we found that communication break-down prevails between the authorities and the stakeholders of the project. Since vital information was not effectively conveyed to the stakeholders, building and sustaining trust are not going to happen. The installation of towers in school compounds project was a good example of communication break-down. The installation was part of a high-profile national initiative to provide equal access to quality teaching and learning by means of using technology in schools. Under this initiative called 1BestariNet, all Malaysian primary and secondary schools are to be equipped with 4G wireless internet connectivity of up to 10 Mbps (Megabits per second) to all urban schools and up to 4 Mbps for rural schools (Sidhu 2012). This internet access is packaged with a virtual learning platform that functions as a web-based learning system for connecting teachers, administrators, students, and parents within one virtual space. It will allow teachers teachers to "assign lessons, tests, and marks virtually, while students can submit homework and view their marks through the Frog VLE. Parents can communicate with schools while school administrators can organize their school calendars and disseminate school notices via the Internet" (1Bestarinet.net 2012). This is clearly a highly ambitious and logistically complex initiative. The required telecommunications infrastructure, designated as the 1BestariNet Radio Integrated System (1BRIS), consists of three main components: a 30-m monopole structure, a radio-frequency panel antenna and microwave dish, and an outdoor cabinet to house essential equipment such as batteries and rectifiers.

The respondents argued that the community was not well-informed about the initiative. Members of parents and teachers' associations (PTAs) claimed that they were not informed about the plan to build telecommunication towers inside the schools. The plans were conveyed to the PTAs through official letters to the schools without any consultations with stakeholders. One respondent noted:

> We received a letter from the school's PTA that a telecommunication tower will be built in this school for the purpose of mobile Internet network… there is no further details about the height, location of the tower, etc. (Parent)

Although it is a big project in terms of cost, key stakeholders and members of PTAs in the community were not engaged. The respondents also claimed that the engagement with the community occurred only after the communication towers have been constructed and after the persistent requests from the PTAs. As the respondents perceived that the communication between the authorities and communities was virtually absent, confusion and anxiousness built up among stakeholders which lead to the distrust in the authorities. One respondent claimed:

> I believe that the schools have no choice … the schools have to accept orders … we are surprised why the tower is so near to the school … why can't they talk to us …' our children's safety is at stake … are they trying to hide anything? (Parent)

(b) **Control and monitoring of the project**

As suggested by Solomon/Flores (2001: 30), when we trust someone to perform a task, it means that we trust his competency, so there is no need to control him. Essentially, you will not control people when you have trust in them. However, you have to control people if you do not trust them.

In the study, we found that there is a strong sense of trust between the authorities and the firm that was awarded with the installation of the towers as evident by the lack of monitoring or controlling. The respondents claimed that the authorities were absent when the construction of the towers took place. As a result, the towers were erected very close to the school buildings or the hostels.

> I saw no security personnel... electric wires all over places ... the contractor is taking advantage of the school ... they don't care about the safety of the children. (Teacher)

By giving full control to the firm to perform the given task, it means that the authorities have fully trusted the firm. As the job was not performed up to the users' expectations, it raises questions about the credibility of the authorities, resulting in the receding of trustworthiness of the authorities.

> The authority has the power to define safety ... but the industry is also powerful ... multi-billion dollar industry ... they can influence the decision. (Parent)

Another parent commented along the same line:

> Once they (the firm) have the tower, they can rent it out ... make more money for them ... who dare to complain about it? This is multi-million dollars company. (Parent)

(c) Competency on the project

Trust in the authorities is related to their capability to perform their responsibility. As suggested by Solomon/Flores (2001: 70), trusting people involves assessing their level of competence, therefore trust is related to competency.

While the respondents agreed about the usefulness of the tower installation in school compounds so as to provide better communication and internet connection, it appears that the deliverables of the project are not up to users' expectations. Respondents argued that the speed of computer network is not enough to support the whole school and there are no significant changes in the learning and teaching process upon the implementation of the project. Instead, the project was only used to support simple task such as attendance and profiling.

> It is very frustrating ... it takes more than one minute just to key-in mark for one student... speed is very slow ... the system is always busy ... what a waste of time. (Teacher)

We found that the promise of delivering an integrated system that can help improve teaching and learning in the schools has not been materialised. Since the authorities have entrusted the firm to deliver the project, success or failure of the project ultimately falls on the authorities. The credit or blame is on the authorities who trust, not the one who is trusted. Therefore, users' dissatisfaction with the project leads to the violation of trust in the authorities; hence it is evidence that competency is related with trust.

> In my opinion, there is nothing wrong with the Government policy in education... but there is no proper evaluation... sometimes they do things without thoroughly studying whether it is suitable in this country or not. (Parent)

(d) **Media coverage of the project**

Media coverage plays an important role in providing information to the public as well as influencing public opinion. Sometimes, the risks are exaggerated as the media compete for public attention and make the issue driven by public sentiment, and not the actual facts (Kahneman 2012: 142).

In our study, we found that media coverage on the 1BestariNet is relatively insignificant despite the substantial cost involved in the project. Issues related to the project were broadcasted in a cable-television news channel (Astro Awani 2014) as well as the local tabloid (Lim 2013). However, no serious media coverage followed up the case; eventually it did not really grab headlines attention.

As a consequence, the respondents turned to various media sources to gather information. Arguably, social media has slightly alleviated the fear although it has not reached to the point that may affect the 1BestariNet project. Nevertheless, the respondents are of the view that the information obtained from the authorities was neither meaningful nor reliable, instead they listened to other sources they thought more reliable.

> I don't think I have seen or heard anything about it from the mainstream or government media ... may be it is not important ... I don't know why. (Parent)

Findings from our study indicate that trust can highly influence people's perception of risks related to telecommunication towers. Although the project is deemed useful, the installation of telecommunication towers to implement the project may cause inconvenience to some parties.

Despite the small number of respondents taking part in our study, which is typical in many qualitative studies, it has successfully managed to capture people's perceptions that may represent others who have the same views. By adopting a qualitative approach, this study is able to gain a better understanding and insights so that ideas for improvement can be generated and implemented.

3.4.2 Trusts Can Influence Risk Perceptions

This study reports similar findings. Firstly, we found that the main concern is the siting of the telecommunication towers that is too close to the school buildings which is perceived to affect the health and safety of school children. Their perception could be less hostile if the location of the towers is far away from the schools. This perception is formed based on the information they thought credible. Although the government agrees that the level of radiofrequency exposure in this country is still safe (Malaysian Communications and Multimedia Commission 2014), the information provided by the government is often perceived less reliable (Hunt et al. 1999; Macoubrie 2006).

One of the ways to improve the credibility of information provided by the government is to engage independent organisation, since reports provided by the industry might be perceived biased as the industry is likely to protect its own interest (Frewer et al. 1996). Therefore, we recommend that reports and publications should be jointly prepared by the industry and independent organisation. We also recommend the authorities to improve the presentation of reports and information on their websites by taking into consideration the differences of users' ability to understand technical issues. In other words, the issues related to radiofrequency on the websites should be more presentable and not too simple or too difficult to understand.

In academics, peer reviewing of research paper is common in order to determine the quality and credibility of the research (Ware 2008). While it is not a perfect tool, we recommend the authorities to do the same with their websites. We suggest peer review exercise where a government agency reviews the website of other government agency in terms of presentation and contents. The exercise may occur among government agencies under the same ministry or among government agencies that have common responsibilities.

Secondly, this study also finds that trust in the government influences the perception of the telecommunication towers as well as the project. Since the 1BestariNet project involved the installation of telecommunication towers inside the schools, it has raised concerns among the community about the risk of radiofrequency. Events surrounding the installation of the towers did not help improve public perception. For example, the firm that operates the project can take advantage of the towers.

It is obvious that the ineffectiveness of communication and lack of engagement with the local communities add further distrust in the authorities. The lack of monitoring by the authorities during the construction of the communication towers has strengthened public distrust in the authorities.

It appears that the perceived risks of the telecommunication towers have been intensified when the deliverables of the project have not met users' expectations. For example, the slow speed affects teaching and learning in the schools. This has contributed to further distrust in the authorities.

Moreover, as mainstream media do not substantially cover or follow the project, alternative media such as social media help spread the news about the project. As the public might have the perception that the authorities are controlling the media, the people are more inclined to pay attention to negative news (Trussler/Soroka 2014). Although information about the project is available from many sources online, building and sustaining public trust are not an easy task once people have made up their judgments.

Building trust involves communication and understanding of the surrounding issues as some issues may require a different approach of communication to different group of communities (Freimuth et al. 2014). Therefore, we recommend the authorities to exercise public relations and communication. With a proactive role of communicating with the public, trust can slowly be built which can relieve uncertainty and tension in the community. We suggest the improvement of the management of public relations. Although there is no guarantee that communication can help change public perception, doing nothing does not help public relations as well (Williams/Olaniran 1998).

It is evident in the study that the respondents perceived that the school children are potentially exposed to health and safety-related risks. Upon further investigation, we have shown that the perception of risks prevailed because there is a lack of trust in the authorities, which in turn affects the perceived benefits of the project. Since there is a lack of trust in the authorities, it affects the sources of reliable information. Information provided by the government is perceived less credible, hence affecting their perceived benefits.

In short, the lack of trust in the authorities adds to greater perception that there is covering-up of the project. Nevertheless, the respondents agree that the project is useful in enhancing the experience of learning and teaching in schools. However, health risks associated with the technology should not be ignored because long term exposure to radiofrequency has not yet been scientifically verified (World Health Organisation 2014).

References

1Bestarinet.net. (2012). *FAQ*. http://btp.moe.gov.my/1bestarinet/article/165.

Astro Awani. (2014). *Guru mempertikai projek 1Bestarinet* (Teachers questioning 1Bestarinet project) [Video file]. Retrieved from http://www.astroawani.com.

Dohle, S., Keller, C., & Siegrist, M. (2012). Mobile communication in the public mind: Insights from free associations related to mobile phone base stations. *Human and Ecological Risk Assessment, 18*, 649–668.

Ferguson, E. (2001). The roles of contextual moderation and personality in relation to the knowledge-risk link in workplace. *Journal of Risk Research, 4*, 323–340.

Freimuth, V. S., Musa, D., Hilyard, K., Quinn, S. C., & Kim, K. (2014). Trust during the early stages of the 2009 H1N1 pandemic. *Journal of Health Communication, 19*, 321–339.

Frewer, L. J., Howard, C., Hedderley, D., & Shepherd, R. (1996). What determines trust in information about food-related risks? *Underlying Psychological Constructs, Risk Analysis, 16* (4), 473–486.

Glaser, B. G., & Strauss, L. A. (1967). *The discovery of grounded theory: Strategies for qualitative research*. Chicago: Aldine Publishing Co.

Hunt, S., Frewer, L. J., & Shepherd, R. (1999). Public trust in sources of information about radiation risks in the UK. *Journal of Risk Research, 2*, 167–180.

Kahneman, D. (2011). *Thinking, fast and slow*. New York: Farrar, Strauss and Giroux.

Lim, I. (2013). *With 1BestariNet, YTL eyes return on concessions scene*. The Malay Mail. Retrieved from http://www.themalaymailonline.com.

Macoubrie, J. (2006). Nanotechnology: Public concerns, reasoning and trust in government. *Public Understanding of Science, 15,* 221–241.

Malaysian Communications and Multimedia Commission. (2014). *Our responsibility.* Retrieved from http://www.skmm.gov.my.

Sidhu, B. K. (2012, April 10). *RM663mil wireless access for 9,924 schools.* The Star. Retrieved from http://www.thestar.com.my/story/?file=%2f2012%2f4%2f10%2fbusiness%2f11075859& sec=.

Siegriest, M., & Cvetkovich, G. (2000). Perceptions of hazard: The role of social trust and knowledge. *Risk Analysis, 20,* 713–719.

Slovic, P. (1987). Perception of risk. *Science, 236,* 280–285.

Solomon, R. C., & Flores, F. (2001). *Building trusts in business, politics, relationships, and life.* New York: Oxford University Press.

Trussler, M., & Soroka, S. (2014). Consumer demand for cynical and negative news frames. *International Journal of Press/Politics, 19,* 360–379.

Uslaner, M. (2007). Trust and risk: Implications for management. In M. Siegrist, T. C. Earl, & H. Gutscher (Eds.), *Trust in cooperative risk management: Uncertainty and scepticism in the public mind.* New York: Earthscan.

Ware, M. (2008). *Peer review: Benefits, perceptions and alternatives.* Publishing Research Consortium, London. Retrieved from http://www.publishingresearch.org.uk/documents/ PRCsummary4Warefinal.pdf.

Williams, D. E., & Olaniran, B. A. (1998). Expanding the crisis planning function: Introducing elements of risk communication to crisis communication practice. *Public Relations Review, 24,* 387–400.

World Health Organization. (2014). *Electromagnetic fields and public health: Mobile phones.* Retrieved from http://www.who.int/news-room/fact-sheets/detail/electromagnetic-fields-and-public-health-mobile-phones.

Chapter 4
Quantitative Findings: Investigating Antecedents of RFEMF Risk Perceptions

A. Noorhidawati

4.1 The Respondents

Upon analysis of the quantitative data, a total of 1,975 respondents took part in the survey. Using the information obtained from the national census of 2010, we found that our sample fairly represented the whole population. Although the distribution of ethnicity did not exactly match with the national census, other demographic findings from the survey fairly reflected the population as a whole. Therefore, based on the descriptive analysis of the sample, misrepresentation of population was not really a big concern. Although it was found that many respondents were not aware of the telecommunication towers installed nearest to their home, it does not mean that the respondents were not aware of the perceived risk of RFEMF.

It was also apparent that the internet was the most powerful source of information and the most preferred media. Therefore, all relevant stakeholders must adopt strategies that are based on the usage of the internet. It meant that people's perception and reservations towards certain issues related to RFEMF may be challenged and countered if information was efficiently channelled and communicated.

Table 4.1 shows the profiles of respondents participating in the survey. The survey was undertaken comprehensively that well-represented the population in Malaysia such as gender, ethnicity, level of education, generation age, marital status, the number of people in a household and the region where they are residing. These socio-demographic profiles are important because they may have significant effect on the perception of risk among the general public.

The distribution of gender in the sample does reflect the population in the country. However, the ethnicity of respondents is not well distributed mainly because the Malays are the majority population; in the range of 45–50% of the population. Nevertheless, the gap between the sample and population is not significantly huge; therefore we do not expect any significant effect on the analysis. In terms of generation age, although the respondents are mostly young, we do not

Y. Kamarulzaman et al., *Public Perceptions of Radiation Effects on Health Risks and Well-Being*, SpringerBriefs in Environment, Security, Development and Peace 33, https://doi.org/10.1007/978-981-32-9894-1_4

Table 4.1 Profile of respondent (n = 1,975)

Profile	Segment	Frequency	Percentage (%)
Gender	Male	839	42.5
	Female	1,136	57.5
Ethnicity	Malay	722	36.6
	Chinese	738	37.4
	Indian	290	14.7
	Others	225	11.4
Generation	Y	939	48.0
	X	791	40.0
	Baby boomers	232	12.0
	Traditionalist	13	1.0
Marital status	Married	972	49.0
	Single	1,003	51.0
Household size	2	319	16.2
	3–5	1,087	55.0
	6–8	457	23.1
	9–11	75	3.8
	12 and above	37	1.9
Residing region	Northern Region	348	17.6
	East Coast Region	208	10.5
	Central Region	922	46.7
	Southern Region	286	14.5
	East Malaysia	186	9.4
	Not Specified	25	1.3
	Total	1,975	100.0

expect the skewed distribution of generation age to severely affect the study. This is because the young generation can greatly influence political decision and policy makers; therefore their voices are important for future generations.

The distribution between married and singles among respondents is nearly even. This indicates specific current social situation in Malaysia and not showing the unintended bias since it is not expected that the ratio of single to married is around 1 in Malaysia. Census data from Department of Statistics Malaysia reported 60% of married and 40% of single population in 2010. In terms of economic background, most of the respondents are living in the house of between three to five people in a household. Therefore, we expect that most of the respondents belong to the middle-class. Most of the respondents are also residing in the central region which is the most populated area in the country. In conclusion, the samples that we managed to get in the study were relatively unbiased and capable of representing the whole population well.

4.2 Perception of Risk by Demographic Profiles

This section discusses the perception of risk according to their demographic profiles and investigates whether there is any significant difference in risk perception between one group and the other.

4.2.1 Gender

Men and women have different views related to the risk of RFEMF as follows. Firstly, men are more inclined to accept telecommunication technology as long as it could improve performance, productivity and effectiveness as long as the benefits are greater than the risk. Secondly, men tended to be more influenced by the views of their family members and friends with regards to the risk of RFEMF. Thirdly, men are less health-conscious than women.

In terms of risk perception of RFEMF, men and women are different in the following aspects. Firstly, women are more anxious and fearful than men when they thought about the risk of RFEMF. Secondly, women are also easily agitated and unreceptive when a new base station of mobile phones is constructed nearby. The results from psychological and risk perception aspects conclude that women are more anxious about RFEMF because they are more health conscious than men, while men are more concerned about the benefits of telecommunication development.

The effects of the risk perception towards their behaviour are as follows. Although women are more anxious towards the risk of RFEMF, they are less hostile than men. This is because women are of the view that the society is highly responsible and each individual should control and reduce the usage of mobile devices. However, risk perception among men are different considering their social role as representative of family members who worry about health effect.

In summary, women tend to be more resourceful, health-conscious and sensitive with regards to the risk of RFEMF but less hostile, while men tend to be less sensitive but may be hostile. However, there is no conclusive evidence to suggest that women are more risk averse than men.

4.2.2 Ethnicity

There are three major ethnics in Malaysia; the Malays, Chinese and Indians. Since Malaysians are identified by their ethnicity and embedded culture, ethnicity may have an effect on respondents' view. It is found that the degree of perception towards RFEMF among ethnics in Malaysia varies. For example, the Malays are more agreeable on the usefulness of technology and they rely more on the views of friends and family members with regards to RFEMF matters. The Malays also believe that media reports have significantly influenced people's view and are more concerned about the value of health.

In terms of perceived risk of RFEMF, the Malays are more anxious and fearful of the risk of RFEMF, while the Chinese express more anger about telecommunication development. Nevertheless, the Malays are more receptive of the government's policy on telecommunication development.

The risk perception of RFEMF affects the behaviour of people of different ethnics differently. However, there is no concrete evidence to suggest that one ethnic is becoming more hostile than the others. For example, the Malays are more agreeable about the role of social responsibility while addressing the RFEMF issue. The Malays are also of the view that the response to the RFEMF is up to the individuals to choose and control. The Malays are more welcoming of the development of telecommunication and they will likely to continue using the RFEMF related devices.

In summary, although there is a significant difference in responses among ethnics with regards to the perception of risk, there is no clear indication that one ethnic will behave differently from others. Although multivariate analysis was employed, confounding factors might affect on this interpretation due to the limited sample size using a cluster analysis.

4.2.3 Age

Different generations age may have different views, therefore the age of respondents is divided into four big groups of generations; generation Y which is between 24 and 38 years of age, generation X which is between 39 and 53 years of age, baby boomers which is between 54 and 72 years of age and traditionalist which is above 73 years of age (Robinson 2017). It is found that there are differences of opinion among generations. For example, the young generation Y is more agreeable on the usefulness of telecommunication technology. They are also of the opinion that the media may exaggerate the risk of RFEMF and are the least concerned about health. However, they are more resourceful to obtain more information and engaged with various ways of communication. Hence, the young generation Y uses more mobile devices.

In terms of risk perception of RFEMF, the older generation of baby boomers is more anxious and highly angry about the risk. On the other hand, the elder generation of the traditionalists is of the opinion that the government had good policies with regards to the risk of RFEMF. Although health-consciousness is the least concern among the younger generation Y, they are more afraid of the RFEMF risk than the other generations.

The effects of RFEMF risk perception towards behaviour among generation age may vary. For example, the young generation Y is more of the view that the society is responsible and it is up to the individuals to take action. Despite the risk, the young generation Y is willing to continue using mobile devices; therefore their actions may be less hostile towards the development of telecommunication towers. However, the older generation of baby boomers believes that individuals should have control, while the older traditionalists are more receptive to the development of telecommunication technology. It means that the generation of baby boomers is more likely to behave more hostile than their elders.

In short, although the younger generation is more accepting of telecommunication technology, they are also afraid of the risk. Despite their anxiousness and anger towards telecommunication development, the older generation thought that the government policy could and should be useful to protect people from health risks. Therefore, opinions among generations are hugely diverging and there is no conclusive evidence to suggest that a particular generation is more risk-averse than the others. However, there is evidence to suggest that the baby boomers generation may behave more hostile towards the development of telecommunication towers.

4.2.4   Level of Education

Education helps people think differently, therefore the level of education may have an effect on people's opinion. The level of education among the respondents is divided into two big categories; less- and well-educated people. The less educated people only acquire education up to secondary school, while the well-educated people have at least a university degree. It is found that there is a slight difference of opinion between the two groups. The well-educated group feels that it is important to embrace technological development and is of the view that the media is not helpful in reporting the risk of RFEMF. However, the less-educated group relies more on its social contacts to influence their perception.

Although the well-educated group is less health-conscious, they are more anxious, afraid and highly irritated about the telecommunication development. Nevertheless, the less-educated group finds that the government policy is fair in protecting the public from the risk of RFEMF.

The effects of risk perception towards the behaviour between the well- and less-educated groups are mixed. For example, while the well-educated group feels that the society should be concerned about the risk, they also believe that they could individually protect the society and have more control by expressing their concern on the telecommunication development. Although they are aware of the risk, the well-educated groups will still continue to use mobile devices due to their benefits. Therefore, it is not certain that the well-educated group will become more hostile towards the building of telecommunication towers.

In summary, it is evident that well-educated people have high perception of RFEMF risk, although health-consciousness is not their top priority. Nevertheless, the well-educated group is also found to be more receptive towards telecommunication technology. Therefore, there is no conclusive evidence to suggest that either the well-educated or less-educated group will show more hostility towards telecommunication towers.

4.2.5 Geographical Locations

People who live in different geographical locations may have different perspectives due to the nature of political affinity or ethnicity influences embedded in a particular location. So, the locations of the respondents' residence may have an effect on their opinions. The geographical locations in the country could be divided into five big regions. The central region is considered the wealthiest and the most populated region that experiences more telecommunication developments than other regions, while the East Malaysia region is considered the least developed region.

It is found that there is a slight difference of opinion among the regions. For example, people in the central region feel that telecommunication development is necessary, while people in southern region are heavily influenced by their social circles with regards to the risk of RFEMF. People in the east coast region are more health conscious, but people in the southern region are more anxious, afraid and highly irritated about the development of telecommunication. On the other hand, people in East Malaysia are easily satisfied with the telecommunication development in accordance with their current needs.

4.3 The Antecedents of RFEMF Risk Perceptions

There are three main factors that affect the perceived risk of RFEMF; psychological, personal and demographic factors. Psychological factors include usefulness of technology, material value, media influence and social influence. Personal factors include perceived responsible parties; health consciousness, preferred media, preferred activities, preferred health information source and usage of RFEMF devices. Demographic factors include gender, ethnicity, education and others as shown in Table 4.2.

While perceived risk as a construct is included into four sub-constructs called anger, anxiety, fearfulness and government related, basically it meant that people's anger with regards to RFEMF risk could be used to indicate perceived risk. Table 4.3 indicates that the government policy related to RFEMF risk had the highest number of significant factors, five, while anxiety had the lowest number of significant factors. It shows that the perceived risk was highly related with the government policy and was less related with people's anxiety.

To investigate factors that influence public perception on RFEMF risk, the following research questions were examined:

Table 4.2 MANOVA results for perceived risks – demographic factors

Demographic (Sig)	Data	Perceived risks			
		Anxiety	Fear	Anger	Government policy related
Gender	Value	0.992			
	Sig	0.21*	0.004*	0.688	0.921
Ethnicity	Value	0.957			
	Sig	0.032*	0.002*	0.016*	0.000*
Age	Value	0.977			
	Sig	0.085	0.416	0.343	0.843
Education	Value	0.973			
	Sig	0.728	0.003*	0.011*	0.576
Income	Value	0.982			
	Sig	0.14	0.196	0.437	0.024*
Status	Value	0.990			
	Sig	0.481	0.205	0.36	0.385
Work category	Value	0.943			
	Sig	0.101	0.15	0.046*	0.042*
Position	Value	0.970			
	Sig	0.396	0.996	0.479	0.035*
Family size	Value	0.986			
	Sig	0.966	0.159	0.487	0.012*
House location	Value	0.984			
	Sig	0.252	0.084	0.1	0.337
Work location	Value	0.987			
	Sig	0.208	0.849	0.701	0.671

4.3.1   What Are the Relationships Between Demographic Profiles and Public Risk Perception Associated with RFEMF?

Based on the MANOVA and regression results, there was a significant relationship between the demographic profile and public risk perception on RFEMF. Both results indicated that ethnicity, gender, level of education and work category statistically influenced the perceived risks as measured by anger, anxiety, fearfulness and government policy related.

Different ethnicities had different views of risk as shown in Table 4.4. However, the results showed that the Malays had high risk perception compared to other ethnicities. Similarly, people with high education and more income had higher risk perception of RFEMF. The Chinese had relatively higher perceived risk (for anger sub-scale, mean = 3.37) towards RFEMF compared to other ethnicities. However, the Malay ethnicity had higher perceived risk for fear sub-scale (mean = 3.40). It indicated that it was important for responsible parties to target different ethnicities

Table 4.3 MANOVA results for perceived risks – personal and psychological factors

Construct	Independent variables	Data	Perceived risks			
			Anxiety	Fear	Anger	Government policy related
Personal factors	Preferred activities	Value	0.989			
		Sig	0.296	0.589	0.765	0.538
	Usage of RFEMF devices	Value	0.968			
		Sig	0.228	0.112	0.091	0.002*
	Preferred media	Value	0.988			
		Sig	0.5	0.497	0.16	0.131
	Preferred health information source	Value	0.978			
		Sig	0.010*	0.043*	0.09	0.141
	Preferred responsible parties	Value	0.983			
		Sig	0.331	0.082	0.626	0.084
	Health conscious	Value	0.951			
		Sig	0.000*	0.104	0.341	0.014*
Psychological factors	Usefulness of technology	Value	0.919			
		Sig	0.000*	0.000*	0.000*	0.179
	Material values	Value	0.954			
		Sig	0.000*	0.003*	0.000*	0.000*
	Social influence	Value	0.816			
		Sig	0.000*	0.000*	0.000*	0.111
	Media influence	Value	0.894			
		Sig	0.000*	0.000*	0.000*	0.003*

*Significant at 0.05 levels (2-tailed)

Table 4.4 Mean scores of different views of risk perception

Dependent variable		Mean	Std. error
Anxiety	Malay	2.988	0.240
	Chinese	3.003	0.239
	Indian	2.834	0.239
	Others	2.804	0.276
Fearfulness	Malay	3.401	0.272
	Chinese	3.296	0.270
	Indian	3.091	0.270
	Others	3.328	0.312
Anger	Malay	3.277	0.232
	Chinese	3.365	0.230
	Indian	3.182	0.231
	Others	3.166	0.266
Government policy	Malay	2.658	0.261
	Chinese	2.320	0.259
	Indian	2.553	0.260
	Others	2.355	0.300

with different and innovative approaches in order to maintain effective communication to reduce perceived risks.

4.3.2   What Are the Relationships Between Personal and Psychological Factors Towards Public Risk Perceptions of RFEMF?

There were a number of personal and psychological factors influencing public risk perceptions of RFEMF; preferred health information sources and health consciousness as shown in the summary of regression in Table 4.5, from the highest to the lowest score.

From the regression coefficients table above, social influence has the highest impact on risk perception, followed by media influence, sources of information and health consciousness. The usefulness of technology has a negative impact on risk perception. This meant that people who are of the view that technology was not really useful have a high risk perception, whereas people who are inclined towards technology have less risk perception of RFEMF. Material values have the weakest contribution towards risk perception. This meant that people did not really put high material value as an indication of risk perception.

The implication of the result showed the strength of the power of social influence, such as word of mouth, either through physical contact or through online social networks. Moreover, as more information related to health topics is passed to the public, it increases the level of perceived risks directly and significantly. The public tends to be confused with overwhelming information in which reliability and accuracy of information are still questionable. Due to this, the public has a tendency to evoke negative feelings towards RFEMF.

Nevertheless, when the public is of the view that the technology is very useful, then it helps reduce the perceived risks on RFEMF. Nowadays, with internet availability anywhere, the viral effect on news spreading is unbelievable. Therefore, practitioners should leverage on the advantage of these influences strategically in reducing the perceived risks level among the public. Moreover, health information related to RFEMF should be shared in transparent, simple and objective manner in

Table 4.5 Contribution of personal and psychological factors towards risk perception

	Significant factor	Unstandardised coefficient
Personal factors	Preferred health information source	0.113
	Health consciousness	0.106
Psychological factors	Social influence	0.428
	Media influence	0.167
	Material values	−0.031
	Usefulness of technology	−0.202

order to provide opportunities to the public to digest and understand it easily. This may mitigate their perceived risk and behaviour.

4.3.3 Summary of Quantitative Findings

There are eight factors affecting perceived risks of RFEMF. Four factors are grouped under personal factors namely, (i) social influence, (ii) media influence, (iii) material values and (iv) usefulness of technology. Another two come under psychological factors, (v) preferred health information source and (vi) health consciousness. The last two are demographic factors, (vii) work category and (viii) ethnicity (Fig. 4.1).

While the risk of RFEMF is recognisable and unavoidable, the relevant stakeholders should be more proactive and committed to communicate and rectify the perception of RFEMF. The debate on the risk of RFEMF will not disappear as long as the general public gets the wrong information from the Internet and other sources of information.

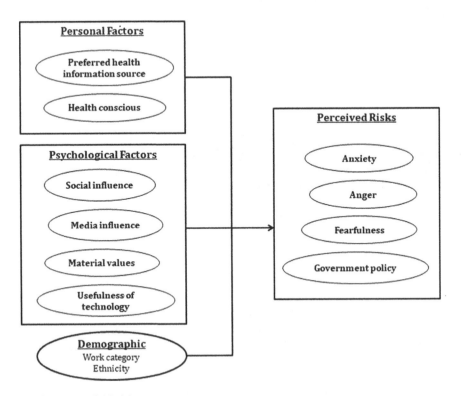

Fig. 4.1 Research Model

For example, recently there were claims that telecommunication towers that were built inside the schools' premises in Perak did not get the consent of the schools' Parents and Teachers' Associations. In other words, local communities were not fully informed and there were further claims that some of the important information was purposely hidden. This will eventually destroy the trust of the general public if no action is taken by the relevant authorities to fully communicate the real truth behind the setting up of the telecommunication towers.

However, based on different levels of perceived risks and the presence of other factors, the public will react differently. If positive messages on RFEMF are in the public mind-set, they will have good attitudes in using the devices and will continue using them. At the same time, they will recommend the benefits of these devices to the friends. Hence, it encourages more users and people will feel safe when using them.

Nevertheless, there are few demographic variables that turned out to be insignificant in affecting the perceived risks towards RFEMF, such as the type of house they live in, physical health conditions and their knowledge about telecommunication towers.

Reference

Robinson, M. (2017). Which generation are you? Retrieved from https://www.careerplanner.com/Career-Articles/Generations.cfm.

Chapter 5
Conclusion and Recommendations

F. D. Yusop

5.1 Summary of Findings

5.1.1 Qualitative Findings

The findings of the qualitative data collections indicated that public acceptance of RFEMF risk was influenced by several factors which are listed below:

- Emotional responses (anger and fear)
- Media influence (media content, availability, usage, word of mouth)
- Behavioural belief and control
- Social responsible consumption
- Health conscious
- Attitude towards development
- Material values
- Subjective norms
- Political influence.

Subsequently, these findings were used as a framework in devising the survey instrument of the quantitative study. These factors supported by other established instruments from literature reviews were further studied in a larger sample size (2,000 samples) through survey questionnaires. As mentioned earlier, the focus of this report is quantitative results, so the summary of qualitative findings is presented in this chapter.

A small segment of the public is worried about the risk of radiofrequency related to the telecommunications tower, although it is not a huge concern compared to other hazards such as nuclear power. Nevertheless, some countries have prepared rules and regulations relating to the telecommunication towers as recommended by the World Health Organization (2014). However, we believe that political biases may influence people's perception as the project is potentially beneficial to the society. Nonetheless, managing and communicating risk perceptions to the public requires numerous psychosocial and technical elements involved in managing risk communication, including the importance of building trust in the authorities, greater enforcement of laws, competency of the authority and serious and honest engagement with the community.

This study also demonstrates that qualitative methodology can be useful to provide detailed explanations that could contribute towards a better understanding on why a segment of the public reacts differently with respect to telecommunications tower. Findings of this study are beneficial to the policy makers and the authorities to ensure that communication of risks is not taken lightly. Managing public perception necessitates appropriate action plans so that public concerns and welfare are taken care of besides building trust between the public and the authorities.

5.1.2 Quantitative Findings

Data from qualitative findings were later tested in quantitative study. From the quantitative findings, it is found that people's perception of risk posed by RFEMF can be grouped into psychological and personal dimensions.

(a) **Psychological dimension**

Psychological factors are related to an individual psychology that drives their perceptions towards RFEMF. There are four factors within psychological dimension: usefulness of technology, material values, social and media influence.

Telecommunication technology, in this study, refers to mobile devices which are seen as an important part of people's lives. If the public perceives the devices as important, they view the RFEMF risks as low because the benefits of the technologies are seen as greater than the risks. Similarly, if they perceive the devices as adding values to their personal and work life, the RFEMF risks are seen low.

Social and media influences also determine public perceptions. Positive social and media influences towards telecommunication towers and devices will moderate the individual's perceptions towards the associated RFEMF risks and vice versa.

(b) **Personal dimension**

Personal dimension refers to individual's health consciousness and his/her preferred health information sources. People who are health conscious are more likely to perceive higher risk posed by RFEMF than those who are not too conscious. Likewise, people who are highly resourceful and use various media to obtain information about RFEMF are more likely to perceive it as low risk.

5.2 Conclusions

In general, people are aware of the risks associated with the use of electronic devices in our daily lives either at the workplace or at home, but their perceptions are influenced by the factors in two dimensions: psychological and personal. Therefore, it is important for policymakers and stakeholders to take the responsibility to respond and react accordingly to increase public awareness of RFEMF.

The risk of exposure to electromagnetic fields is inevitable due to the advancement of technology and widespread use of the internet via Wi-Fi. Nearly 2,000 people participated in the survey and their demographic profiles shown in the descriptive analysis above more or less reflect the Malaysian population. Due to certain assumptions of this study (target respondents must have a certain level of knowledge about RFEMF), the target sample is expected to be skewed toward Generation Y.

Although people are generally aware of the risk of RFEMF, there is no strong evidence to show that they understand it well and correctly. Therefore, we observed that there is a need to continue educating the public to generate greater awareness levels. At the same time, there is no direct impact of the risk of RFEMF to their health. This is because there are no cases of health problems that can be attributed directly to the presence of RFEMF which might due to inability to reach the respondents with health problems caused by RFEMF that indicate a possibility of Type 2 error.

It is also found that people regularly use the internet and the influence of social media networks is recognised. The relevant stakeholders should therefore develop their strategies of communicating about the usage of the internet as the media.

5.3 Recommendations

Based on the discussion of the findings above, there are several recommendations that could be executed immediately by authorities and policymakers.

5.3.1 Communication of Risks

The Government should aim at increasing and improving the management of communication in order to reduce bias and prejudice. This is also important in order to educate the public on the risks involved with RFEMF. This can be done by providing quarterly newsletter to schools.

5.3.2 Building Trust

In order to improve public trust in the authority, the Government should increase engagement with all relevant stakeholders by participating in public forums and debates.

5.3.3 Public Relation Exercises

As a way to show that the authority cares, the Government should increase public presence by participating in television talk shows.

5.3.4 Information Asymmetry

In order to encourage and build a well-informed society, the Government should improve dissemination of quality information (offline and online) by providing complete information and links for further details.

5.3.5 Managing Reputation

Public perception must be managed and improved. So, the authority must convince the public that they are responsible, playing their parts and protecting the rights of the people. This can be done by continuously showing or making the public aware of the enforcement steps that have been taken or executed.

5.3.6 Online Presence

The authorities should improve the management of information online since the most affected people are in the urban and suburban areas. The Government should provide links to credible audio/video information.

5.4 Implications of Study

These are the implications of the study.

5.4.1 Understanding Target Audiences

In order to gauge what suitable strategy should be used to easily reach them out, the relevant stakeholders need to know well the target public demographic profiles. For example, by understanding the profile of ethnicity, income, family size etc., it is advised to communicate the message about RFEMF strategically.

5.4.2 Choosing Appropriate Media Channels

Stakeholders can use the appropriate preferred responsible parties, preferred activity etc. to reach the public. For example, an activity like awareness campaign from the responsible parties by conducting a preferred activity helps increase public knowledge, awareness and confidence level.

5.4.3 Designing Communication Content

The most important part will be the content of what is to be communicated. For reducing the risks perception on RFEMF, the suggestion is to use simple, memorable language for communication. Scientifically, terms in RFEMF sounded like a stranger to the public. This is mainly because the public does not use such scientific language in daily topic sharing. Therefore, to be part of the community members, it is advisable to use a common language for effective communication.

5.4.4 The Role of Authorities and Policymakers

Government authorities may serve as an organisation responsible for information clarification, responding to public concerns related to RFEMF and regularly providing updates on the meetings/conferences related to RFEMF issues. This is to engage with the public and indirectly it helps increase public awareness towards RFEMF. If the public is well-informed on RFEMF and gets up-to-date and reliable information, then the perceived risks related to RFEMF will be minimised.

The authorities may serve as a 'channel' for the public to raise their concerns (complaints and complimentary) related to issues of RFEMF. Currently, there is no proper and formal organisation/body which can uphold this better than government authorities. With this implementation, it helps increase public engagement as well. Good and quality customers' services are important in better performing this responsibility.

Proactively organise some forums and question and answer sessions through preferred media such as newspaper and TV channel. This can be done regularly on a quarterly or bi-annually basis. This helps gauge better public responses proactively. Thus, the public will think that there is some person and/ or organisation that really cares about them and are sincere and committed to listen and help them.

5.5 Limitations and Future Research

The study was limited by two factors, a skewed sample and the unidentified variables. Overall, the sample was skewed towards generation Y which has the largest representation in the samples. The presence of unidentified variables led to their exclusion from further analyses.

To address these limitations, future research should eliminate biases in samples by using stratified random or systematic sampling. In order to improve sample representativeness, sample size may be significantly increased from 2,000 to 3,000. The quality of data could be improved by providing small rewards or incentives with each questionnaire. This could overcome the challenge in approaching respondents as well. Finally, an in-depth study is vital in order to identify and include other unidentified variables in the model. This is important in order to generate better reflection on public perceptions.

Reference

World Health Organization. (2014). *Electromagnetic fields and public health: Mobile phones.* Retrieved from http://www.who.int/news-room/fact-sheets/detail/electromagnetic-fields-and-public-health-mobile-phones.

Appendices

Appendix A: Photos of Telecommunication Tower Within School's Premise

These two photos show examples of the close proximity between the telecommunication tower and the school building at two different schools. The presence of towers is common in most public schools. *Sources* Authors

Appendix B: Photos of Public Protests Pertaining
to the Fear of Telecommunication Towers' Hazards

These two photos show examples of public protests that took place at different locations and times as reported in the media. The main reason for the protests was that they deemed the construction of telecommunication towers was too close to their home which could be the source of health hazards. This type of news coverage has been sporadically reported by the media as early as 2005 onwards. *Sources* www.utusan.com.my (published on 21 May 2014) and www.sinarharian.com. my (published on 3 April 2016).

Appendix C: Survey Questionnaires

4. Please circle the best answer that fits you for each item. If you feel strongly negative about it, please tick number '1'. If you feel strongly positive, please tick number '5'. If your opinion is in between, please choose among numbers 2, 3, or 4.

H5) For me, continuous development on telecommunication is:						
AT1	Harmful	1	2	3	4	5	Beneficial
AT2	Unpleasant	1	2	3	4	5	Pleasant
AT3	Bad	1	2	3	4	5	Good
AT4	Worthless	1	2	3	4	5	Valuable
AT5	Un-enjoyable	1	2	3	4	5	Enjoyable

H6) Overall, I feel the current government's policy on telecommunication development is:						
P1	Unsatisfactory	1	2	3	4	5	Satisfactory
P2	Static	1	2	3	4	5	Growing
P3	Bad	1	2	3	4	5	Good
P4	Worthless	1	2	3	4	5	Valuable
P5	Declining	1	2	3	4	5	Improving
P6	Restricted	1	2	3	4	5	Flexible
P7	Biased	1	2	3	4	5	Fair

5. For the activities listed below, please indicate _how often you engaged_ in during your free time.

How often do you get involved in the following activities during your free time?	Never	Seldom	Sometimes	Often	Always
AC1	Internet browsing and interaction (e.g. email, chatting, Facebook etc.)	1	2	3	4	5
AC2	Telephone interaction (e.g. phone calls, video calls, SMS, MMS etc.)	1	2	3	4	5
AC3	Socialising (e.g. attend gathering, reunion, association, informal meetings etc.)	1	2	3	4	5
AC4	Knowledge seeking (e.g. attend trainings, seminar, lectures etc.)	1	2	3	4	5
AC5	Social service work (e.g. orphanage, hospital, school, CSR etc.)	1	2	3	4	5
AC6	Reading (e.g. books, newspaper, magazines etc.)	1	2	3	4	5
AC7	Sports and games (e.g. gym, fitness class, indoor, outdoor etc.)	1	2	3	4	5
AC8	Entertainment (e.g. music, TV, cinema, karaoke etc.)	1	2	3	4	5
AC9	Others (pls. specify)	1	2	3	4	5

RF-EMF risk in this study is defined as the possible radiation risk produced by all wireless communication devices and cell phone transmission towers. -5-

6. Please _state your media preference in the space provided and circle the level of engagement_ of the stated media.

State your most preferred media for each category below and the level engagement....	Never	Seldom	Sometimes	Often	Always
PM1	Newspaper (e.g. Berita Harian, The Star etc.)	1	2	3	4	5
PM2	Magazine (e.g. Remaja, Cleo, Wanita etc.)	1	2	3	4	5
PM3	TV Channels (e.g. Astro, TV3 etc.)	1	2	3	4	5
PM4	Radio (e.g. Sinar, Hitz Fm, BFM etc.)	1	2	3	4	5
PM5	Social Media (e.g. Facebook, Twitter etc.)	1	2	3	4	5

7. Please _circle your level of usage_ of the following devices.

How frequent do you use the following devices	Never	Seldom	Sometimes	Often	Always
U1	Cell Phone / Smart Phone	1	2	3	4	5
U2	Cordless Phone	1	2	3	4	5
U3	Radio	1	2	3	4	5
U4	Television	1	2	3	4	5
U5	Microwave Oven	1	2	3	4	5
U6	Personal Computer / Laptop	1	2	3	4	5
U7	iPad / Tablet / PDA	1	2	3	4	5
U8	Digital Alarm Clock	1	2	3	4	5
U9	Wi-fi Routers	1	2	3	4	5
U10	Bluetooth Devices	1	2	3	4	5
U11	Video game consoles	1	2	3	4	5
U12	Medical / laboratory equipment	1	2	3	4	5
U13	Smart meter (measure energy consumption)	1	2	3	4	5
U14	Wireless Devices (e.g. modem, mouse, alarm, printer, baby monitor etc.)	1	2	3	4	5

RF-EMF risk in this study is defined as the possible radiation risk produced by all wireless communication devices and cell phone transmission towers. -6-

---

SECTION B

Instruction: Please fill in the space provided or tick (/) the answer that BEST describes you.

Gender
- ☐ Male
- ☐ Female

Ethnicity
- ☐ Malay
- ☐ Chinese
- ☐ Indian
- ☐ Others, pls. specify_____

Age (at last birthday)
- ☐ Below 20 years old
- ☐ 20 - 30 years old
- ☐ 31 - 40 years old
- ☐ 41 - 50 years old
- ☐ 51 - 60 years old
- ☐ 61 - 70 years old
- ☐ 71 - 80 years old
- ☐ Above 80 years old

Educational Background
- ☐ Primary
- ☐ Secondary
- ☐ Undergraduate Degree
- ☐ Postgraduate Degree
- ☐ Professional Degree
- ☐ Others, pls. specify_____

Monthly Household Income
- ☐ Below RM 2,000
- ☐ RM 2,001 - RM 5,000
- ☐ RM 5,001 - RM 8,000
- ☐ RM 8,001 - RM 11,000
- ☐ Above RM 11,000

Marital Status
- ☐ Married with children
- ☐ Married without children
- ☐ Single parent
- ☐ Single
- ☐ Others, pls. specify_____

Work Category
- ☐ Accounting / Finance
- ☐ Admin / Human Resources
- ☐ Arts / Media / Communications
- ☐ Building / Construction
- ☐ Computer / Information Technology
- ☐ Education / Training
- ☐ Engineering
- ☐ Healthcare
- ☐ Hotel / Restaurant
- ☐ Manufacturing
- ☐ Sales / Marketing
- ☐ Sciences
- ☐ Services
- ☐ Others, pls. specify_____

Current Position
- ☐ Top Management
- ☐ Middle Management
- ☐ Executive / Officer
- ☐ First line Management
- ☐ Team Leader
- ☐ Operational Worker
- ☐ Students
- ☐ Housewife
- ☐ Pensioner
- ☐ Others, pls. specify_____

RF-EMF risk in this study is defined as the possible radiation risk produced by all wireless communication devices and cell phone transmission towers. -7-

---

No of Household (including you)
- ☐ 2
- ☐ 3 - 5
- ☐ 6 - 8
- ☐ 9 - 11
- ☐ 12 and above

Please state your nearest town. (e.g. Shah Alam, Petaling Jaya, Sungai Petani, Muar, etc.)
_____

Type of housing you live in
- ☐ Apartment / Condominium
- ☐ Bungalow
- ☐ Flats
- ☐ Linked-house / Terrace
- ☐ Semi Detached
- ☐ Townhouse / Duplex
- ☐ Others, pls. specify_____

Nearest telecommunication / cell phone tower from your house (approx.).
- ☐ Don't know
- ☐ Below 500 metres
- ☐ 501 to 1000 metres
- ☐ 1001 to 1500 metres
- ☐ 1501 to 2000 metres
- ☐ Above 2000 metres

Are you surrounded by other homes with Wi-fi installed?
- ☐ Yes
  If Yes, pls. specify approximately how many houses have Wi-fi
- ☐ No
- ☐ Not Sure

Nearest telecommunication / cell phone tower from your place of work (approx.).
- ☐ Don't know
- ☐ Below 500 metres
- ☐ 500 to 1000 metres
- ☐ 1001 to 1500 metres
- ☐ 1501 to 2000 metres
- ☐ Above 2000 metres

How would you rate your physical health?
- ☐ Excellent / Very Healthy
- ☐ Good / Healthy
- ☐ Fair / Somewhat Healthy
- ☐ Poor / Unhealthy
- ☐ Bad / Very Unhealthy

When did you last visit your doctor and what was the reason?
- ☐ In the past 3 months
- ☐ Between 3 and 6 months ago
- ☐ Between 6 months and 12 months ago
- ☐ More than 12 months ago

Reason:_____

Thank you for your time and effort participating in this survey.

RF-EMF risk in this study is defined as the possible radiation risk produced by all wireless communication devices and cell phone transmission towers. -8-

Printed in the United States
By Bookmasters